MEMOIRS
of the
American Mathematical Society

ISSN 0065-9266 (print)
ISSN 1947-6221 (online)

Number 1331

Global Dynamics Above the Ground State Energy for the Combined Power-Type Nonlinear Schrödinger Equations with Energy-Critical Growth at Low Frequencies

Takafumi Akahori
Slim Ibrahim
Hiroaki Kikuchi
Hayato Nawa

July 2021 • Volume 272 • Number 1331 (first of 7 numbers)

AMERICAN MATHEMATICAL SOCIETY

Library of Congress Cataloging-in-Publication Data

Cataloging-in-Publication Data has been applied for by the AMS.
See http://www.loc.gov/publish/cip/.
DOI: https://doi.org/10.1090/memo/1331

Memoirs of the American Mathematical Society

This journal is devoted entirely to research in pure and applied mathematics.

Subscription information. Beginning with the January 2010 issue, *Memoirs* is accessible from `www.ams.org/journals`. The 2021 subscription begins with volume 269 and consists of six mailings, each containing one or more numbers. Subscription prices for 2021 are as follows: for paper delivery, US$1085 list, US$868 institutional member; for electronic delivery, US$955 list, US$764 institutional member. Upon request, subscribers to paper delivery of this journal are also entitled to receive electronic delivery. If ordering the paper version, add US$20 for delivery within the United States; US$80 for outside the United States. Subscription renewals are subject to late fees. See `www.ams.org/help-faq` for more journal subscription information. Each number may be ordered separately; *please specify number* when ordering an individual number.

Back number information. For back issues see `www.ams.org/backvols`.

Subscriptions and orders should be addressed to the American Mathematical Society, P. O. Box 845904, Boston, MA 02284-5904 USA. *All orders must be accompanied by payment.* Other correspondence should be addressed to 201 Charles Street, Providence, RI 02904-2213 USA.

Memoirs of the American Mathematical Society (ISSN 0065-9266 (print); 1947-6221 (online)) is published bimonthly (each volume consisting usually of more than one number) by the American Mathematical Society at 201 Charles Street, Providence, RI 02904-2213 USA. Periodicals postage paid at Providence, RI. Postmaster: Send address changes to Memoirs, American Mathematical Society, 201 Charles Street, Providence, RI 02904-2213 USA.

This publication is indexed in *Mathematical Reviews®*, *Zentralblatt MATH*, *Science Citation Index®*, *Science Citation Index™-Expanded*, *ISI Alerting Services℠*, *SciSearch®*, *Research Alert®*, *CompuMath Citation Index®*, *Current Contents®/Physical, Chemical & Earth Sciences.*
This publication is archived in *Portico* and *CLOCKSS*.
Printed in the United States of America.

∞ The paper used in this book is acid-free and falls within the guidelines established to ensure permanence and durability.
Visit the AMS home page at `https://www.ams.org/`

10 9 8 7 6 5 4 3 2 1 26 25 24 23 22 21

Contents

Chapter 1. Introduction 1

Chapter 2. Properties of Ground State 11

Chapter 3. Proof of Theorem 1.0.1 17

Chapter 4. Decomposition around ground state 21

Chapter 5. Ejection lemma 39

Chapter 6. Modified distance function 51

Chapter 7. One-pass theorem 57

Chapter 8. Proof of Theorem 3.0.1 71

Appendix A. Existence of ground state 109

Appendix B. Fundamental properties of the linearized operators 115

Appendix C. Inequalities for the radial functions 119

Appendix D. Small-data theory 121

Appendix E. Long-time perturbation theory 123

Appendix F. Table of notation 127

Bibliography 129

Abstract

We consider the combined power-type nonlinear Schrödinger equations with energy-critical growth, and study the solutions slightly above the ground state threshold at low frequencies, so that we obtain a so-called nine-set theory developed by Nakanishi and Schlag.

Received by the editor December 19, 2015, and, in revised form, February 7, 2018.

Article electronically published on September 21, 2021.

DOI: https://doi.org/10.1090/memo/1331

2020 *Mathematics Subject Classification.* Primary .

The first author was partially supported by the Grant-in-Aid for Young Scientists (B) # 25800077 of JSPS. The second author's research was partially supported by NSERC Discovery grant # 371637-2019. The third author was partially supported by the Grant-in-Aid for Young Scientists (B) # 00612277 of JSPS. The last author was partially supported by the Grant-in-Aid for Scientific Research (B) 23340030 of JSPS and the Grant-in-Aid for Challenging Exploratory Research 23654052 of JSPS..

CHAPTER 1

Introduction

We consider the following nonlinear Schrödinger equation:

$$\text{(NLS)} \qquad i\frac{\partial \psi}{\partial t} + \Delta\psi + |\psi|^{p-1}\psi + |\psi|^{\frac{4}{d-2}}\psi = 0,$$

where $\psi = \psi(x,t)$ is a complex-valued function on $\mathbb{R}^d \times \mathbb{R}$ ($d \geq 3$), Δ is the Laplace operator on $\mathbb{R}^d$ and p satisfies that

$$2 + \frac{4}{d} < p+1 < 2^* := 2 + \frac{4}{d-2}. \tag{1.1}$$

We denote the mass and the Hamiltonian of (NLS) by $\mathcal{M}$ and $\mathcal{H}$, respectively, that is,

$$\mathcal{M}(u) := \frac{1}{2}\|u\|_{L^2}^2, \tag{1.2}$$

$$\mathcal{H}(u) := \frac{1}{2}\|\nabla u\|_{L^2}^2 - \frac{1}{p+1}\|u\|_{L^{p+1}}^{p+1} - \frac{1}{2^*}\|u\|_{L^{2^*}}^{2^*}. \tag{1.3}$$

The Cauchy problem for (NLS) is locally well-posed in H^1 (see, e.g., Proposition 3.1 in [**29**]), and the mass and the Hamiltonian are conserved quantities for the flow defined by (NLS). Furthermore, for any solution ψ of finite variance, we have the "virial identity":

$$\frac{d^2}{dt^2}\int_{\mathbb{R}^d} |x|^2|\psi(x,t)|^2\,dx = 8\mathcal{K}(\psi(t)), \tag{1.4}$$

where

$$\mathcal{K}(u) := \|\nabla u\|_{L^2}^2 - \frac{d(p-1)}{2(p+1)}\|u\|_{L^{p+1}}^{p+1} - \|u\|_{L^{2^*}}^{2^*}. \tag{1.5}$$

It is easy to see that

$$\mathcal{H}(u) > \frac{1}{2}\mathcal{K}(u). \tag{1.6}$$

A standing wave of (NLS) with frequency ω is a solution to (NLS) of the form $e^{it\omega}u$, so that u solves the following semilinear elliptic equation:

$$\omega u - \Delta u - |u|^{p-1}u - |u|^{\frac{4}{d-2}}u = 0, \qquad u \in H^1(\mathbb{R}^d)\setminus\{0\}. \tag{1.7}$$

Moreover, a ground state of (1.7) is a solution to (1.7) of the minimal action, where by the action, we mean the functional $\mathcal{S}_\omega$ defined by

$$\mathcal{S}_\omega := \omega\mathcal{M} + \mathcal{H}. \tag{1.8}$$

In [**2**], the same authors showed the existence of a ground state via the variational problem

$$m_\omega := \inf\{\mathcal{S}_\omega(u)\colon u \in H^1(\mathbb{R}^d)\setminus\{0\},\ \mathcal{K}(u)=0\}. \tag{1.9}$$

More precisely, we proved that for any $d \ge 4$, any p with $2+\frac{4}{d} < p+1 < 2^*$ and any $\omega > 0$, the variational problem associated with m_ω has a minimizer, and any minimizer for m_ω becomes a ground state of (1.7). Moreover, we see from Proposition 1.2 in [**2**] that for any $d \ge 3$, any p with $2+\frac{4}{d} < p+1 < 2^*$ and any $\omega > 0$,

$$\begin{aligned} m_\omega &= \inf\{\mathcal{I}_\omega(u)\colon u \in H^1(\mathbb{R}^d)\setminus\{0\},\ \mathcal{K}(u)\le 0\} \\ &= \inf\{\mathcal{J}_\omega(u)\colon u \in H^1(\mathbb{R}^d)\setminus\{0\},\ \mathcal{K}(u)\le 0\}, \end{aligned} \tag{1.10}$$

where

$$\mathcal{I}_\omega(u) := \mathcal{S}_\omega(u) - \frac{2}{d(p-1)}\mathcal{K}(u), \tag{1.11}$$

$$\mathcal{J}_\omega(u) := \mathcal{S}_\omega(u) - \frac{1}{2}\mathcal{K}(u). \tag{1.12}$$

It is worthwhile noting that $\mathcal{I}_\omega$ and $\mathcal{J}_\omega$ do not contain the L^{p+1}-norm and $\dot{H}^1$-norm, respectively: Precisely,

$$\mathcal{I}_\omega(u) = \frac{\omega}{2}\|u\|_{L^2}^2 + \frac{p-1-\frac{4}{d}}{2(p-1)}\|\nabla u\|_{L^2}^2 + \frac{2^*-(p+1)}{2^*(p-1)}\|u\|_{L^{2^*}}^{2^*}, \tag{1.13}$$

$$\mathcal{J}_\omega(u) = \frac{\omega}{2}\|u\|_{L^2}^2 + \frac{d(p-1-\frac{4}{d})}{4(p+1)}\|u\|_{L^{p+1}}^{p+1} + \frac{1}{d}\|u\|_{L^{2^*}}^{2^*}. \tag{1.14}$$

REMARK 1.0.1. *The authors have to say that Theorem 1.2 in* [**2**] *contains a mistake. Actually, if $d=3$ and $3<p<5$, then for any $\omega>0$, we can prove the existence of ground state of* (1.7) *(cf. Theorem 1 in* [**32**]*). Moreover, we find from Proposition* A.0.1 *in Section* A *that if $d=3$ and $1<p<5$, then there exists $\omega_3>0$ such that for any $0<\omega<\omega_3$, the equation* (1.7) *admits at least one ground state.*

Our aim in this paper is to study the behavior of solutions to (NLS) slightly above the "ground state threshold at low frequencies", in the spirit of Nakanishi and Schlag [**24, 25**]. To this end, we need detailed information about ground states of (1.7) . We will see that for any $d \ge 3$ and $2+\frac{4}{d} < p+1 < 2^*$, there exists $\omega_1 > 0$ such that for any $0<\omega<\omega_1$, a positive radial ground state Φ_ω of (1.7) exists and unique (see Proposition 2.0.1 and (i) of Proposition 2.0.4). Furthermore, the ground state Φ_ω satisfies that $\frac{d}{d\omega}\mathcal{M}(\Phi_\omega) < 0$ on $(0,\omega_1)$ (see (iii) of Proposition 2.0.4), so that the inverse of the mapping $\alpha \in (0,\omega_1) \to \mathcal{M}(\Phi_\alpha) \in (\mathcal{M}(\Phi_{\omega_1}),\infty)$ exists (we will see that $\lim_{\omega\to 0}\mathcal{M}(\Phi_\omega)=\infty$ in (2.9)).

Throughout this paper, we use the symbol Φ_ω to denote the positive radial ground state of (1.7). Moreover, $\mathcal{O}(\Phi_\omega)$ denotes the orbit

$$\mathcal{O}(\Phi_\omega) := \{e^{i\theta}\Phi_\omega\colon \theta\in\mathbb{R}\}. \tag{1.15}$$

Now, we state our main result.

THEOREM 1.0.1. *Assume either $d=3$ and $3 \le p < 5$, or $d \ge 4$ and $1+\frac{4}{d-1} < p < 2^*-1$. Then, there exists $\omega_* > 0$ such that for any $\omega \in (0,\omega_*)$, there exists a positive function $\varepsilon_\omega \colon [0,\infty) \to (0,\infty)$ with the following property: Set*

$$\widetilde{PW}_\omega := \{u \in H^1(\mathbb{R}^d) \colon \mathcal{S}_\omega(u) < m_\omega + \varepsilon_\omega(\mathcal{M}(u))\}. \tag{1.16}$$

Then, any radial solution ψ starting from $\widetilde{PW}_\omega$ exhibits one of the following scenarios: In the statement below, $\alpha > 0$ denotes the constant such that $\mathcal{M}(\psi) = \mathcal{M}(\Phi_\alpha)$.
(i) *Scattering both forward and backward in time;*
(ii) *Finite time blowup both forward and backward in time;*
(iii) *Scattering forward in time, and finite time blowup backward in time;*
(iv) *Finite time blowup forward in time, and scattering backward in time;*
(v) *Trapped by $\mathcal{O}(\Phi_\alpha)$ forward in time, and scattering backward in time;*
(vi) *Scattering forward in time, and trapped by $\mathcal{O}(\Phi_\alpha)$ backward in time;*
(vii) *Trapped by $\mathcal{O}(\Phi_\alpha)$ forward in time, and finite time blowup backward in time;*
(viii) *Finite time blowup forward in time, and trapped by $\mathcal{O}(\Phi_\alpha)$ backward in time;*
(ix) *Trapped by $\mathcal{O}(\Phi_\alpha)$ both forward and backward in time.*
Here, "scattering forward in time" means that the maximal lifespan of a solution ψ is infinity and there exists $\phi \in H^1(\mathbb{R}^d)$ such that

$$\lim_{t\to\infty} \left\| \psi(t) - e^{it\Delta}\phi \right\|_{H^1} = 0; \tag{1.17}$$

"blowup forward in time " means that the maximal lifespan of a solution is finite; and "trapped by $\mathcal{O}(\Phi_\alpha)$ forward in time" means that the maximal lifespan of a solution is infinity and the solution stays in some neighborhood of $\mathcal{O}(\Phi_\alpha)$ in $H^1(\mathbb{R}^d)$ after some time. The terms corresponding to "backward in time" are used in a similar manner.

REMARK 1.0.2. *We continue our study in anticipation of validity of Theorem 1.1 for $d \ge 3$ and $1+\frac{4}{d} \le p < 2^*-1$. We describe main obstructions to the extension:*
(1) *The double critical case (namely, $p = 1+\frac{4}{d}$) has a special difficulty that the same proof as* (iii) *of Proposition* 2.0.4 *cannot apply to derive the property* (2.11). *In fact, the property* (2.11) *fails for the equation* (1.22) *with $p = 1+\frac{4}{d}$. Moreover, in the double critical case $p=1+\frac{4}{d}$, the scaling-exponent s_p (see* (1.42)*) vanishes ($s_{1+\frac{4}{d}} = 0$), which requires some modifications in our argument, especially the proof of existence of unstable mode (Proposition* 4.0.1*).*
(2) *The restriction on the exponent p in Theorem* 1.0.1 *comes from the "one-pass theorem" (see Theorem* 7.0.1 *and Remark* 7.0.1*).*

We will undertake the extension of Theorem 1.0.1 *including the double critical case in a forthcoming paper as well as the high-frequency case $\omega \gg 1$.*

We give a proof of Theorem 1.0.1 in Section 3 below. In particular, we show how to construct the function ε_ω (see (3.8)).

The equation (NLS) reminds us of the following ones

$$i\frac{\partial \psi}{\partial t} + \Delta\psi + |\psi|^{p-1}\psi = 0, \tag{1.18}$$

$$i\frac{\partial \psi}{\partial t} + \Delta\psi + |\psi|^{\frac{4}{d-2}}\psi = 0. \tag{1.19}$$

These equations are invariant under the following scalings, respectively:

$$\psi(x,t) \mapsto \lambda^{-\frac{2}{p-1}}\psi(\frac{x}{\lambda},\frac{t}{\lambda^2}), \tag{1.20}$$

$$\psi(x,t) \mapsto \lambda^{-\frac{d-2}{2}}\psi(\frac{x}{\lambda},\frac{t}{\lambda^2}). \tag{1.21}$$

On the other hand, there is no scaling which leaves (NLS) invariant.

The stationary problems corresponding to (1.18) and (1.19) are respectively

$$\omega v - \Delta v - |v|^{p-1}v = 0, \tag{1.22}$$

$$\Delta v + |v|^{\frac{4}{d-2}}v = 0. \tag{1.23}$$

Any solution v to (1.22) gives rise to a standing wave $e^{i\omega t}v$ to (1.18), and any solution to (1.23) solves the equation (1.19). Here, we introduce the "scaling-operator" T_ω to be that for any function f on $\mathbb{R}^d$,

$$T_\omega f(x) := \omega^{-\frac{1}{p-1}} f(\frac{x}{\sqrt{\omega}}). \tag{1.24}$$

Then, putting $u := T_\omega v$, we see that u solves

$$u - \Delta u - |u|^{p-1}u = 0. \tag{1.25}$$

It is well known that the equation (1.25) has a unique positive solution up to translations which is radially symmetric with respect to some point. We denote the positive solution symmetric about the origin by U. On the other hand, the equation (1.23) possesses a solution of the form

$$W(x) := \left(\frac{\sqrt{d(d-2)}}{1+|x|^2} \right)^{\frac{d-2}{2}}. \tag{1.26}$$

Here, a positive C^2-solution of (1.23) is unique up to scalings and translations (see Corollary 8.2 in [**8**]). Following Brézis-Nirenberg [**7**], we introduce the variational value

$$\sigma := \inf\left\{ \|\nabla u\|_{L^2}^2 : u \in \dot{H}^1(\mathbb{R}^d),\ \|u\|_{L^{2^*}} = 1 \right\}. \tag{1.27}$$

Then, it is known that $u = W/\|W\|_{L^{2^*}}$ is a minimizer of the variational problem associated with σ and

$$\sigma^{\frac{d}{2}} = \|\nabla W\|_{L^2}^2 = \|W\|_{L^{2^*}}^{2^*}. \tag{1.28}$$

We also introduce the variational value

$$m^\ddagger := \inf\left\{ \mathcal{I}^\ddagger(u) \colon u \in \dot{H}^1(\mathbb{R}^d) \setminus \{0\},\ \mathcal{K}^\ddagger(u) \le 0 \right\}, \tag{1.29}$$

where

$$\mathcal{I}^\ddagger(u) := \mathcal{H}^\ddagger(u) - \frac{2}{d(p-1)}\mathcal{K}^\ddagger(u) \tag{1.30}$$

with

$$\mathcal{H}^\ddagger(u) := \frac{1}{2}\|\nabla u\|_{L^2}^2 - \frac{1}{2^*}\|u\|_{L^{2^*}}^{2^*}, \tag{1.31}$$

$$\mathcal{K}^\ddagger(u) := \|\nabla u\|_{L^2}^2 - \|u\|_{L^{2^*}}^{2^*}. \tag{1.32}$$

Note that

$$\mathcal{I}^{\ddagger}(u)=\frac{d(p-1)-4}{2d(p-1)}\|\nabla u\|_{L^2}^2+\frac{4-(d-2)(p-1)}{2d(p-1)}\|u\|_{L^{2^*}}^{2^*}. \tag{1.33}$$

In particular, $\mathcal{I}^{\ddagger}(u)\ge 0$ for $1+\frac{4}{d}\le p\le 2^*-1$.

We can verify (see Appendix A) that

$$m^{\ddagger}=\mathcal{I}^{\ddagger}(W)=\mathcal{H}^{\ddagger}(W)=\frac{1}{d}\sigma^{\frac{d}{2}}. \tag{1.34}$$

Furthermore, we proved in [**2**] that for any $d\ge 4$, any $1+\frac{4}{d}<p<2^*-1$ and any $\omega>0$,

$$0<m_{\omega}<\frac{1}{d}\sigma^{\frac{d}{2}}. \tag{1.35}$$

We can also prove (1.35) for $d=3$ if $\omega>0$ is sufficiently small (see Lemma A.0.4).

Now, we go back to the equation (1.7). If u is a solution to (1.7), then $v:=T_{\omega}u$ satisfies that

$$v-\Delta v-|v|^{p-1}v-\omega^{\frac{2^*-(p+1)}{p-1}}|v|^{\frac{4}{d-2}}v=0. \tag{1.36}$$

We should mention that if ω is small enough, it is expected that (1.7) has properties similar to (1.25). Indeed, this is the heart of our analysis.

We briefly review results for (1.18) and (1.19) which are used in this paper.

In [**18**], Kenig and Merle studied the equation (1.19) in the dimensions $d=3,4,5$, and proved the scattering of radial solutions starting from the set

$$PW_{+}^{\ddagger}:=\left\{u\in H^1(\mathbb{R}^d)\colon \mathcal{H}^{\ddagger}(u)<\mathcal{H}^{\ddagger}(W),\ \|\nabla u\|_{L^2}^2<\|\nabla W\|_{L^2}^2\right\}. \tag{1.37}$$

Their result is extended to the higher dimensional cases by Killip and Visan [**21**]. We summarize these results:

THEOREM 1.0.2 ([**18, 21**]). *Assume $d\ge 3$. Then, the set $PW_{+}^{\ddagger}$ is invariant under the flow defined by* (1.19)*. Furthermore, any non-trivial radial solution ψ to* (1.19) *starting from $PW_{+}^{\ddagger}$ exists globally in time, and satisfies that*

$$\|\nabla\psi\|_{St(\mathbb{R})}<\infty, \tag{1.38}$$

$$\mathcal{H}^{\ddagger}(\psi)\ge\frac{1}{2}\inf_{t\in\mathbb{R}}\mathcal{K}^{\ddagger}(\psi(t))>0, \tag{1.39}$$

where $St(\mathbb{R}):=L^{\infty}(\mathbb{R},L^2(\mathbb{R}^d))\cap L^2(\mathbb{R},L^{2^*}(\mathbb{R}^d))$.

The result for (1.18) corresponding to Theorem 1.0.2 is derived by Duyckaerts, Holmer and Roudenko [**11, 14**] for $d=p=3$. Furthermore, the first and the fourth authors extended their result to the general dimensions $d\ge 1$ and the powers p satisfying (1.1) in [**1**]; the result says that we have either the scattering or the blowup, if the solutions start from the set

$$PW^{\dagger}:=\{u\in H^1(\mathbb{R}^d)\setminus\{0\}\colon \mathcal{H}^{\dagger}(u)<\Big(\frac{\mathcal{M}(U)}{\mathcal{M}(u)}\Big)^{\frac{1-s_p}{s_p}}\mathcal{H}^{\dagger}(U)\}, \tag{1.40}$$

where

$$\mathcal{H}^{\dagger}(u):=\frac{1}{2}\|\nabla u\|_{L^2}^2-\frac{1}{p+1}\|u\|_{L^{p+1}}^{p+1}, \tag{1.41}$$

and s_p is the "scaling-exponent" given by

$$s_p := \frac{d}{2} - \frac{2}{p-1}. \tag{1.42}$$

The condition (1.1) on p implies that $0 < s_p < 1$. Furthermore, we can classify the behavior of the solutions by the sign of the functional $\mathcal{K}^\dagger$ defined by

$$\mathcal{K}^\dagger(u) := \|\nabla u\|_{L^2}^2 - \frac{d(p-1)}{2(p+1)}\|u\|_{L^{p+1}}^{p+1}. \tag{1.43}$$

Note here that for any $\omega > 0$, the replacement of U with $T_\omega U$ in (1.40) leaves $PW^\dagger$ unchanged. Moreover, introducing the set $PW_\omega^\dagger$ as

$$PW_\omega^\dagger := \{u \in H^1(\mathbb{R}^d) \setminus \{0\} \colon \mathcal{S}_\omega^\dagger(u) < \mathcal{S}_\omega^\dagger(U)\}, \tag{1.44}$$

we can express $PW^\dagger$ as the union of $PW_\omega^\dagger$ over all $\omega > 0$:

$$PW^\dagger = \bigcup_{\omega>0} PW_\omega^\dagger. \tag{1.45}$$

Here, $\mathcal{S}_\omega^\dagger$ denotes the action for (1.22), namely,

$$\mathcal{S}_\omega^\dagger(u) := \omega\mathcal{M}(u) + \mathcal{H}^\dagger(u). \tag{1.46}$$

In [**25**], Nakanishi and Schlag considered the equation (1.18) in the case $d = p = 3$. They developed a method to analyze the behavior of (radial) solutions starting from the set

$$PW^{\dagger,\varepsilon} := \{u \in H^1(\mathbb{R}^3) \setminus \{0\} \colon \mathcal{H}^\dagger(u) < \left(\frac{\mathcal{M}(U)}{\mathcal{M}(u)}\right)(\mathcal{H}^\dagger(U) + \varepsilon)\} \tag{1.47}$$

for some $\varepsilon > 0$. Clearly, the set $PW^{\dagger,\varepsilon}$ is an enlargement of $PW^\dagger$. In particular, they proved that there exists $\varepsilon > 0$ such that all solutions starting from $PW^{\dagger,\varepsilon}$ exhibit one of the same nine scenarios as Theorem 1.0.1 above. This result motivated our study.

We note that the way to define $PW^\dagger$ (see (1.40)) is based on the scaling-invariant nature of U (cf. [**1**]). Due to lack of such a scaling property, it is not appropriate to define the corresponding "potential well" for our equation (NLS) by simply replacing $\mathcal{H}^\dagger$ and U with $\mathcal{H}$ and Φ_ω, respectively in $PW^\dagger$. Instead, from the viewpoint of (1.45), we consider the frequency-wise "potential well" PW_ω defined by

$$PW_\omega := \{u \in H^1(\mathbb{R}^d) \colon \mathcal{S}_\omega(u) < m_\omega\}. \tag{1.48}$$

This set is closely related to a variational nature of Φ_ω. Indeed, the definition of m_ω (see (1.9)) implies that if $u \in H^1(\mathbb{R}^d) \setminus \{0\}$ and $\mathcal{K}(u) = 0$, then $\mathcal{S}_\omega(u) \geq m_\omega$, so that we can split PW_ω into three parts according to the sign of the functional $\mathcal{K}$:

$$PW_\omega = PW_{\omega,+} \cup \{0\} \cup PW_{\omega,-}, \tag{1.49}$$

where

$$PW_{\omega,+} := \{u \in H^1(\mathbb{R}^d) \colon \mathcal{S}_\omega(u) < m_\omega,\ \mathcal{K}(u) > 0\}, \tag{1.50}$$

$$PW_{\omega,-} := \{u \in H^1(\mathbb{R}^d) \colon \mathcal{S}_\omega(u) < m_\omega,\ \mathcal{K}(u) < 0\}. \tag{1.51}$$

The following theorem follows from the results in [**2**, **3**] together with the existence theorem of ground state in $\mathbb{R}^3$ (see Proposition A.0.1 and Proposition A.0.4, and [**32**]).[1]

THEOREM 1.0.3. *Let $\omega > 0$. Assume $d \geq 3$ and $1 + \frac{4}{d} < p < 2^* - 1$. Furthermore, assume $\omega \in (0, \omega_3)$ if $d = 3$, where ω_3 is the frequency given in Proposition A.0.1. Then, the following hold: $PW_{\omega,+}$ and $PW_{\omega,-}$ are invariant under the flow defined by* (NLS)*; and*
(i) *Any radial solution ψ to* (NLS) *starting from $PW_{\omega,+}$ satisfies*

$$\|\langle\nabla\rangle\psi\|_{St(\mathbb{R})} < \infty, \tag{1.52}$$

$$\mathcal{H}(\psi) \geq \frac{1}{2} \inf_{t\in\mathbb{R}} \mathcal{K}(\psi(t)) > 0, \tag{1.53}$$

where $St(\mathbb{R}) := L^\infty(\mathbb{R}, L^2(\mathbb{R}^d)) \cap L^2(\mathbb{R}, L^{2^*}(\mathbb{R}^d))$. In particular, ψ scatters both forward and backward in time.
(ii) *Any radial solution ψ to* (NLS) *starting from $PW_{\omega,-}$ satisfies*

$$\sup_{t\in I_{\max}(\psi)} \mathcal{K}(\psi(t)) < 0, \tag{1.54}$$

and blows up in a finite time both forward and backward in time, where $I_{\max}(\psi)$ denotes the maximal existence-interval of ψ.

Note that Theorem 1.0.1 is an extension of Theorem 1.0.3.

This paper is organized as follows. In Section 2, we give fundamental properties of ground state of (1.7). In Section 3, we prove Theorem 1.0.1. The main ingredient of the proof is Theorem 3.0.1. We assign preliminaries for the proof of Theorem 3.0.1 to Section 4 through Section 7: In Section 4, we think about the decomposition of solution around the ground state in the same viewpoint of Nakanishi and Schlag [**25**]. In Section 5, we prove the "ejection lemma" for the equation (NLS). In Section 6, we introduce the distance function used in [**25**], and derive some variational lemmas. In Section 7, we prove the "one-pass theorem". In Section 8, we finally prove Theorem 3.0.1. This paper also has appended sections: In Section A, we prove the existence of ground state of (1.7) in three dimensions. In Section B, we state fundamental properties of linearized operator around a ground state of (1.7). In Section C, we give two well-known inequalities for radial functions. In Section D, we record a small-data theory, and in Section E we give a long-time perturbation theory in a general setting.

Notation. Besides the notation introduced so far, we use the following notation (see also the table of notation in Section F):

(i) We use $\partial_\omega\Phi_\omega$ to denote the derivative of the positive radial grand state Φ_ω of (1.7) with respect to ω (see Proposition 2.0.4 below for the differentiability).

(ii) The daggered symbols are related to (1.25):

$$L_+^\dagger := 1 - \Delta - pU^{p-1}, \tag{1.55}$$

$$L_-^\dagger := 1 - \Delta - U^{p-1}. \tag{1.56}$$

[1]The papers [**2**, **3**] dealt with the case where $d \geq 4$ and $2 + \frac{4}{d} < p+1 < 2^*$ only. However, we can easily verify that the arguments in [**2**, **3**] work well for the case where $d = 3$ and $1 + \frac{4}{3} < p < 5$ as long as there exists a ground state Q_ω with $m_\omega = \mathcal{S}_\omega(Q_\omega)$.

(iii) Functionals with tilde are related to the rescaled equation (1.36):

$$\widetilde{\mathcal{K}}_\omega(u) := \|\nabla u\|_{L^2}^2 - \frac{d(p-1)}{2(p+1)}\|u\|_{L^{p+1}}^{p+1} - \omega^{\frac{2^*-(p+1)}{p-1}}\|u\|_{L^{2^*}}^{2^*}, \tag{1.57}$$

$$\widetilde{\mathcal{I}}_\omega(u) := \frac{1}{2}\|u\|_{L^2}^2 + \frac{s_p}{d}\|\nabla u\|_{L^2}^2 + \omega^{\frac{2^*-(p+1)}{p-1}}\frac{1-s_p}{d}\|u\|_{L^{2^*}}^{2^*}, \tag{1.58}$$

$$\widetilde{L}_{\omega,+} := 1 - \Delta - p\big(T_\omega\Phi_\omega\big)^{p-1} - \omega^{\frac{2^*-(p+1)}{p-1}}(2^*-1)\big(T_\omega\Phi_\omega\big)^{\frac{4}{d-2}}, \tag{1.59}$$

$$\widetilde{L}_{\omega,-} := 1 - \Delta - \big(T_\omega\Phi_\omega\big)^{p-1} - \omega^{\frac{2^*-(p+1)}{p-1}}\big(T_\omega\Phi_\omega\big)^{\frac{4}{d-2}}. \tag{1.60}$$

(iv) We use $(\cdot,\cdot)_{L^2}$ to denote the inner product in $L^2(\mathbb{R}^d)$:

$$(u,v)_{L^2} := \int_{\mathbb{R}^d} u(x)\overline{v(x)}\,dx.$$

We also use $L^2_{real}(\mathbb{R}^d)$ to denote the real Hilbert space of complex-valued functions in $L^2(\mathbb{R}^d)$ which is equipped with the inner product

$$(u,v)_{L^2_{real}} := \Re\int_{\mathbb{R}^d} u(x)\overline{v(x)}\,dx.$$

Furthermore, $H^1_{real}(\mathbb{R}^d)$ denotes the real Hilbert space of functions in $H^1(\mathbb{R}^d)$ equipped with the inner product

$$(u,v)_{H^1_{real}} := (u,v)_{L^2_{real}} + (\nabla u, \nabla v)_{L^2_{real}}.$$

(v) We use $\langle v,u\rangle_{H^{-1},H^1}$ to denote the duality pair of $u \in H^1_{real}(\mathbb{R}^d)$ and $v \in H^{-1}_{real}(\mathbb{R}^d)$:

$$\langle v,u\rangle_{H^{-1},H^1} := ((1-\Delta)^{-\frac{1}{2}}v, (1-\Delta)^{\frac{1}{2}}u)_{L^2_{real}}.$$

(vi) Let I be an interval, and let $1+\frac{4}{d} \le q \le 2^*-1$. Then, we introduce Strichartz-type spaces on I as

$$St(I) := L_t^\infty L_x^2(I) \cap L_t^2 L_x^{2^*}(I), \tag{1.61}$$

$$V_{q+1}(I) := L_t^{\frac{(d+2)(q-1)}{2}} L_x^{\frac{2d(d+2)(q-1)}{d(d+2)(q-1)-8}}(I), \tag{1.62}$$

$$W_{q+1}(I) := L_{t,x}^{\frac{(d+2)(q-1)}{2}}(I). \tag{1.63}$$

Note here that Sobolev's embedding shows $|\nabla|^{-s_q}V_{q+1}(I) \hookrightarrow W_{q+1}(I)$. Moreover, by Strichartz' estimate, we mean the following estimate: for any appropriate space-time function u, any $t_0 \in I$ and any pair $(q,r) \in [2,2^*] \times [2,\infty]$ with $\frac{2}{r} = d(\frac{1}{2}-\frac{1}{q})$,

$$\|u\|_{St(I)} \lesssim \|u(t_0)\|_{L^2} + \|i\frac{\partial u}{\partial t} + \Delta u\|_{L_t^{r'}L_x^{q'}(I)}, \tag{1.64}$$

where q' and r' denote the Hölder conjugates of q and r respectively.

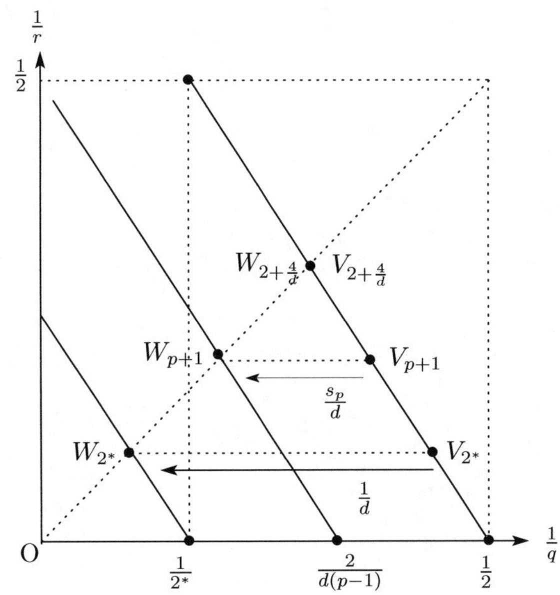

FIGURE 1. Strichartz-type spaces

(vii) For a space-time function u, we define

$$F[u] := |u|^{p-1}u + |u|^{\frac{4}{d-2}}u, \quad F^{\dagger}[u] := |u|^{p-1}u, \quad F^{\ddagger}[u] := |u|^{\frac{4}{d-2}}u, \tag{1.65}$$

$$e[u] := i\frac{\partial u}{\partial t} + \Delta u + F[u], \quad e^{\ddagger}[u] := i\frac{\partial u}{\partial t} + \Delta u + F^{\ddagger}[u]. \tag{1.66}$$

CHAPTER 2

Properties of Ground State

In this section, we give several fundamental properties of ground state of (1.7). As mentioned in Section 1, for any $d \geq 4$, any $1+\frac{4}{d} < p < 2^*-1$ and any $\omega > 0$, the equation (1.7) admits at least one ground state Q_ω satisfying $m_\omega = \mathcal{S}_\omega(Q_\omega)$ (see Proposition 1.1 and Theorem 1.1 in [**2**]). When $d=3$, it follows from Proposition A.0.1 and Proposition A.0.4 in Appendix A that for any $1+\frac{4}{3} < p < 5$, there exists $\omega_3 > 0$ such that for any $0 < \omega < \omega_3$, the equation (1.7) admits a ground state Q_ω satisfying $m_\omega = \mathcal{S}_\omega(Q_\omega)$.

The following proposition tells us that any ground state of (1.7) is essentially positive and radial:

PROPOSITION 2.0.1. *Assume $d \geq 3$ and $1+\frac{4}{d} < p < 2^*-1$. Let $\omega > 0$, and let Q_ω be a ground state of* (1.7)*. Then, there exist $\theta \in \mathbb{R}$, $y \in \mathbb{R}^d$ and a positive radial ground state Φ_ω of* (1.7) *such that $\Phi_\omega \in C^2(\mathbb{R}^d) \cap H^1(\mathbb{R}^d)$, $\frac{x}{|x|}\cdot \nabla \Phi_\omega < 0$ and $Q_\omega = e^{i\theta}\Phi_\omega(\cdot - y)$.*

PROOF OF PROPOSITION 2.0.1. Let Q_ω be a ground state of (1.7). Then, using (1.10), we can verify that $|Q_\omega|$ is also a ground state of (1.7). We also see that $|Q_\omega| \in W^{2,q}_{loc}(\mathbb{R}^d)$ for any $2 \leq q < \infty$ (see Theorem 2.3 in [**6**]). Furthermore, the Schauder theory shows that $Q_\omega \in C^2(\mathbb{R}^d)$ and $\lim_{|x|\to\infty} |Q_\omega(x)| = 0$. On the other hand, it follows from Theorem 9.10 in [**23**] that for any compact set K in $\mathbb{R}^d$, there exists a constant C depending only on K and ω such that for any $x \in K$,

$$|Q_\omega(x)| \geq C \int_K |Q_\omega(y)|\, dy. \tag{2.1}$$

Hence, $|Q_\omega|$ is a positive ground state of (1.7). We see from the result of Gidas, Ni and Nirenberg [**12**] that there exist a positive radial function Φ_ω with $\frac{x}{|x|}\cdot\nabla\Phi_\omega(x) < 0$ for any $x \in \mathbb{R}^d \setminus \{0\}$, and $y \in \mathbb{R}^d$, such that $|Q_\omega| = \Phi_\omega(\cdot - y)$. We introduce the sign of a complex-valued function u, denoted by $\operatorname{sgn} u$, as

$$\operatorname{sgn} u := \frac{\overline{u}}{|u|}. \tag{2.2}$$

Then, we have

$$\operatorname{sgn} \overline{Q_\omega} := \frac{Q_\omega}{|Q_\omega|}, \tag{2.3}$$

so that $Q_\omega = |Q_\omega| \operatorname{sgn} \overline{Q_\omega}$. We shall show that $\operatorname{sgn} \overline{Q_\omega}$ is constant, which together with $|\operatorname{sgn} \overline{Q_\omega}| \equiv 1$ completes the proof. Since

$$\Re\big[\operatorname{sgn}\overline{Q_\omega}\nabla \operatorname{sgn} Q_\omega(x)\big] = \Re\Big[\frac{Q_\omega}{|Q_\omega|}\nabla\Big(\frac{\overline{Q_\omega}}{|Q_\omega|}\Big)\Big] = \frac{\Re[Q_\omega \nabla\overline{Q_\omega}] - |Q_\omega|\nabla|Q_\omega|}{|Q_\omega|^2} = 0, \tag{2.4}$$

we have

$$
(2.5)\quad
\begin{aligned}
\|\nabla Q_\omega\|_{L^2}^2 &= \left\| \nabla \left(\, |Q_\omega| \mathrm{sgn}\, \overline{Q_\omega} \, \right) \right\|_{L^2}^2 \\
&= \int_{\mathbb{R}^d} \left|\nabla |Q_\omega(x)|\right|^2 dx + \int_{\mathbb{R}^d} \left|Q_\omega(x)\right|^2 \left|\nabla \mathrm{sgn}\, \overline{Q_\omega}(x)\right|^2 dx \\
&\quad + \int_{\mathbb{R}^d} \nabla \left|Q_\omega(x)\right|^2 \Re\left[\mathrm{sgn}\, \overline{Q_\omega} \nabla \mathrm{sgn}\, Q_\omega(x)\right] dx \\
&= \int_{\mathbb{R}^d} \left|\nabla |Q_\omega(x)|\right|^2 dx + \int_{\mathbb{R}^d} \left|Q_\omega(x)\right|^2 \left|\nabla \mathrm{sgn}\, \overline{Q_\omega}(x)\right|^2 dx.
\end{aligned}
$$

Hence, if $\nabla \mathrm{sgn}\, \overline{Q_\omega} \not\equiv 0$, then we would have

$$
(2.6)\qquad m_\omega = \mathcal{S}_\omega(Q_\omega) > \mathcal{S}_\omega(|Q_\omega|) = \mathcal{S}_\omega(\Phi_\omega) = m_\omega.
$$

However, this is a contradiction. Thus, $\mathrm{sgn}\, \overline{Q_\omega}$ must be constant. □

Next, we give a decay property of a positive ground state.

PROPOSITION 2.0.2. *Assume $d \ge 3$ and $1 + \frac{4}{d} < p < 2^* - 1$. Let $\omega > 0$, and let $\Phi_\omega \in C^2(\mathbb{R}^d) \cap H^1(\mathbb{R}^d)$ be a positive radial ground state of* (1.7)*. Then, there exist positive constants $C(\omega) > 0$ and $\delta(\omega) > 0$ depending on ω such that*

$$
(2.7)\qquad |\Phi_\omega(x)| + |\nabla \Phi_\omega(x)| + |\Delta \Phi_\omega(x)| \le C(\omega) e^{-\delta(\omega)|x|}
$$

for any $x \in \mathbb{R}^d$. In particular, $\Phi_\omega \in H^2(\mathbb{R}^d)$.

PROOF OF PROPOSITION 2.0.2. We can prove the proposition in a way similar to [**5**]. □

The following convergence result justifies the intuition that the equation (1.7) looks like (1.25) for a sufficiently small ω.

PROPOSITION 2.0.3. *Assume $d \ge 3$ and $1 + \frac{4}{d} < p < 2^* - 1$. Let $\omega > 0$, and let Φ_ω be a positive radial ground state of* (1.7)*. Moreover, let U be the unique positive radial ground state of* (1.25)*. Then, we have*

$$
(2.8)\qquad \lim_{\omega \downarrow 0} \|T_\omega \Phi_\omega - U\|_{H^1} = 0.
$$

PROOF OF PROPOSITION 2.0.3. We can prove the proposition in a way similar to [**16**]. □

REMARK 2.0.1. *We see from Proposition* 2.0.3 *that for any $1 \le q \le 2^* - 1$,*

$$
(2.9)\qquad \lim_{\omega \downarrow 0} \omega^{s_p - \frac{q-1}{p-1}} \|\Phi_\omega\|_{L^{q+1}}^{q+1} = \|U\|_{L^{q+1}}^{q+1}
$$

and

$$
(2.10)\qquad \lim_{\omega \downarrow 0} \omega^{s_p - 1} \|\nabla \Phi_\omega\|_{L^2}^2 = \|\nabla U\|_{L^2}^2.
$$

PROPOSITION 2.0.4. *Assume that $d \ge 3$ and $1 + \frac{4}{d} < p < 2^* - 1$. Then, there exists $\omega_1 > 0$ with the following properties:*
(i) *For any $\omega \in (0, \omega_1)$, a positive radial ground state of* (1.7) *uniquely exists.*

(ii) *The mapping* $\omega \in (0, \omega_1) \mapsto \Phi_\omega \in H^1(\mathbb{R}^d)$ *is continuously differentiable.*
(iii) *For any* $\omega \in (0, \omega_1)$,

$$\frac{d}{d\omega}\mathcal{M}(\Phi_\omega) = \left(\Phi_\omega, \partial_\omega \Phi_\omega\right)_{L^2_{real}} < 0. \tag{2.11}$$

Here, Φ_ω *denotes the unique positive radial ground state of* (1.7).

In order to prove the claim (i) in Proposition 2.0.4, we need the following result (see, e.g., Theorem 0.2 in [**17**], [**26**] and Proposition 2.8 in [**31**]):

LEMMA 2.0.5. *Assume either* $d = 1, 2$ *and* $1 < p < \infty$, *or* $d \ge 3$ *and* $1 < p < 2^* - 1$. *Then, we have*

$$\operatorname{Ker} L_+^\dagger = \operatorname{span}\{\partial_1 U, \dots, \partial_d U\}, \tag{2.12}$$

where $L_+^\dagger$ *is the operator defined by* (1.55), *and* $\partial_1 U, \dots, \partial_d U$ *denote the partial derivatives of* U.

Now, we give a proof of the claim (i):

PROOF OF (i) IN PROPOSITION 2.0.4. It suffices to prove the uniqueness of the ground state. Suppose for contradiction that for any $k \in \mathbb{N}$, there exists $\omega_k \in (0, \frac{1}{k})$ such that the equation (1.7) has two different positive radial ground states, say U_{ω_k} and V_{ω_k}. Put $\widetilde{U}_k := T_{\omega_k} U_{\omega_k}$, $\widetilde{V}_k := T_{\omega_k} V_{\omega_k}$ and

$$\widetilde{u}_k := \frac{\widetilde{U}_k - \widetilde{V}_k}{\|\widetilde{U}_k - \widetilde{V}_k\|_{H^1}}. \tag{2.13}$$

Then, $\widetilde{u}_k$ satisfies the equation

$$\widetilde{u}_k - \Delta \widetilde{u}_k - \frac{\widetilde{U}_k^p - \widetilde{V}_k^p}{\|\widetilde{U}_k - \widetilde{V}_k\|_{H^1}} - \omega_k^{\frac{2^*-(p+1)}{p-1}} \frac{\widetilde{U}_k^{2^*-1} - \widetilde{V}_k^{2^*-1}}{\|\widetilde{U}_k - \widetilde{V}_k\|_{H^1}} = 0. \tag{2.14}$$

Moreover, it follows from $\widetilde{u}_k \in H^1_{rad}(\mathbb{R}^d)$ and $\|\widetilde{u}_k\|_{H^1} = 1$ for any $k \in \mathbb{N}$ that there exists a radial function $\widetilde{u} \in H^1(\mathbb{R}^d)$ such that, passing to a subsequence, we have

$$\lim_{k\to\infty} \widetilde{u}_k = \widetilde{u} \qquad \text{weakly in } H^1(\mathbb{R}^d), \tag{2.15}$$

and for any $2 < q < 2^*$,

$$\lim_{k\to\infty} \widetilde{u}_k = \widetilde{u} \qquad \text{strongly in } L^q(\mathbb{R}^d). \tag{2.16}$$

We shall show that $\widetilde{u} \equiv 0$. First note that the fundamental theorem of calculus together with Proposition 2.0.3 and (2.16) implies that for any $\phi \in C_c^\infty(\mathbb{R}^d)$,

$$\begin{aligned}
&\lim_{k\to\infty} \left\langle \frac{\widetilde{U}_k^p - \widetilde{V}_k^p}{\|\widetilde{U}_k - \widetilde{V}_k\|_{H^1}}, \phi \right\rangle_{H^{-1},H^1} \\
&= \lim_{k\to\infty} p \int_0^1 \left\langle \{\widetilde{V}_k + \theta(\widetilde{U}_k - \widetilde{V}_k)\}^{p-1} \widetilde{u}_k, \phi \right\rangle_{H^{-1},H^1} d\theta \\
&= p \int_{\mathbb{R}^d} U^{p-1} \widetilde{u} \phi \, dx.
\end{aligned} \tag{2.17}$$

We can also verify that for a given function $\phi \in C_c^\infty(\mathbb{R}^d)$,

$$\lim_{k\to\infty} \omega_k^{\frac{2^*-(p+1)}{p-1}} \left\langle \frac{\widetilde{U}_k^{2^*-1} - \widetilde{V}_k^{2^*-1}}{\|\widetilde{U}_k - \widetilde{V}_k\|_{H^1}}, \phi \right\rangle_{H^{-1},H^1} = 0. \tag{2.18}$$

Combining (2.14) with (2.15), (2.17) and (2.18), we find that for any $\phi \in C_c^\infty(\mathbb{R}^d)$,

$$\langle\, L_+^\dagger \widetilde{u},\, \phi\, \rangle_{H^{-1},H^1} = \langle\, \widetilde{u} - \Delta\widetilde{u} - pU^{p-1}\widetilde{u},\, \phi\, \rangle_{H^{-1},H^1} = 0. \tag{2.19}$$

Furthermore, since $\widetilde{u}$ is a radial H^1-solution of $L_+^\dagger \widetilde{u} = 0$, we find that $\widetilde{u} \in H^2_{rad}(\mathbb{R}^d)$. Hence, it follows from Lemma 2.0.5 (there is no radial function in the kernel of $L_+^\dagger$) that $\widetilde{u} \equiv 0$.

Now, multiplying the equation (2.14) by $\widetilde{u}_k$ and then integrating the resulting equation over $\mathbb{R}^d$, we obtain

$$\|\widetilde{u}_k\|_{H^1}^2 = \int_{\mathbb{R}^d} \frac{\widetilde{U}_k^p - \widetilde{V}_k^p}{\|\widetilde{U}_k - \widetilde{V}_k\|_{H^1}} \widetilde{u}_k\, dx + \omega_k^{\frac{2^*-(p+1)}{p-1}} \int_{\mathbb{R}^d} \frac{\widetilde{U}_k^{2^*-1} - \widetilde{V}_k^{2^*-1}}{\|\widetilde{U}_k - \widetilde{V}_k\|_{H^1}} \widetilde{u}_k\, dx. \tag{2.20}$$

Then, it follows from (2.16) and $\widetilde{u} \equiv 0$ that the right-hand side of (2.20) tends to 0 as $k \to \infty$, whereas the left-hand side is identically 1 (see (2.13)). This is a contradiction. Thus, we have completed the proof of the claim (i) in Proposition 2.0.4. □

Next, we mention how to prove the claim (ii) in Proposition 2.0.4.

PROOF OF (ii) IN PROPOSITION 2.0.4. The claim (ii) follows from the result by Shatah and Strauss [**27**] (see also [**20**]). □

Finally, we move on to the proof of the claim (iii) in Proposition 2.0.4. Let us notice that the claim (iii) is an analogy to that

$$\frac{d}{d\omega}\mathcal{M}(T_{\omega^{-1}}U) = \frac{1}{2}\frac{d}{d\omega}(\omega^{-s_p}\|U\|_{L^2}^2) = -\frac{s_p}{2}\omega^{-s_p-1}\|U\|_{L^2}^2 < 0. \tag{2.21}$$

Hence, it is convenient to introduce a function ΛU defined by

$$\Lambda U(x) := \partial_\omega\{T_{\omega^{-1}}U(x)\}\big|_{\omega=1} = \frac{1}{p-1}U(x) + \frac{1}{2}x\cdot\nabla U(x). \tag{2.22}$$

Then, we can verify that ΛU obeys

$$L_+^\dagger \Lambda U = -U. \tag{2.23}$$

Differentiation of the both sides of the equation (1.7) with respect to ω yields

$$\omega\partial_\omega\Phi_\omega - \Delta\partial_\omega\Phi_\omega - p\Phi_\omega^{p-1}\partial_\omega\Phi_\omega - (2^*-1)\Phi_\omega^{\frac{4}{d-2}}\partial_\omega\Phi_\omega = -\Phi_\omega. \tag{2.24}$$

Furthermore, we see from (2.24) that

$$\widetilde{L}_{\omega,+}(\omega T_\omega\partial_\omega\Phi_\omega) = -T_\omega\Phi_\omega, \tag{2.25}$$

where $\widetilde{L}_{\omega,+}$ is the operator defined by (1.59). We state a property of the operator $\widetilde{L}_{\omega,+}$.

LEMMA 2.0.6. *Assume $d \ge 3$ and $1+\frac{4}{d} < p < 2^*-1$. Then, there exist $\omega_0 > 0$ and $C > 0$ such that for any $\omega \in (0,\omega_0)$ and any $f \in H^2_{rad}(\mathbb{R}^d)$,*

$$\|\widetilde{L}_{\omega,+}f\|_{L^2} \ge C\|f\|_{H^1}. \tag{2.26}$$

PROOF OF LEMMA 2.0.6. Suppose for contradiction that for any $k \in \mathbb{N}$, there exist $\omega_k \in (0,\frac{1}{k})$ and $f_k \in H^2_{rad}(\mathbb{R}^d)$ with $\|f_k\|_{H^1} = 1$ such that

$$\|\widetilde{L}_{\omega_k,+}f_k\|_{L^2} < \frac{1}{k}. \tag{2.27}$$

Then, we can take $f_\infty \in H^1_{rad}(\mathbb{R}^d)$ such that, passing to some subsequence,

$$\lim_{k\to\infty} f_k = f_\infty \qquad \text{weakly in } H^1(\mathbb{R}^d), \tag{2.28}$$

and for any $2 < q < 2^*$,

$$\lim_{k\to\infty} f_k = f_\infty \qquad \text{strongly in } L^q(\mathbb{R}^d). \tag{2.29}$$

Furthermore, for any $\phi \in C_c^\infty(\mathbb{R}^d)$, we have

$$\begin{aligned} &\left|\langle L_+^\dagger f_\infty, \phi\rangle_{H^{-1},H^1}\right| \\ &\le \left|\left\langle \widetilde{L}_{\omega_k,+} f_k, \phi \right\rangle_{H^{-1},H^1}\right| + \left|\left\langle \widetilde{L}_{\omega_k,+}(f_\infty - f_k), \phi \right\rangle_{H^{-1},H^1}\right| \\ &\quad + p\left|\left\langle\{(T_{\omega_k}\Phi_{\omega_k})^{p-1} - U^{p-1}\}f_\infty,\, \phi\right\rangle_{H^{-1},H^1}\right| \\ &\quad + \omega_k^{\frac{2^*-(p+1)}{p-1}}(2^*-1)\left|\left\langle(T_{\omega_k}\Phi_{\omega_k})^{\frac{4}{d-2}} f_\infty,\, \phi\right\rangle_{H^{-1},H^1}\right|. \end{aligned} \tag{2.30}$$

This together with the hypothesis (2.27), (2.28), (2.29), Proposition 2.0.3 and $\lim_{k\to\infty}\omega_k = 0$ shows that $L_+^\dagger f_\infty = 0$ in the distribution sense. Since $(\operatorname{Ker} L_+^\dagger)\big|_{H^1_{rad}} = \{0\}$ (see Lemma 2.0.5), we conclude that $f_\infty = 0$. Furthermore, this fact $f_\infty = 0$ together with $\|f_k\|_{H^1} = 1$, (2.29) and $\lim_{k\to\infty}\omega_k = 0$ gives us that

$$\lim_{k\to\infty} (\widetilde{L}_{\omega_k,+} f_k, f_k)_{L^2} = 1. \tag{2.31}$$

However, it follows from the hypothesis (2.27) that

$$\lim_{k\to\infty} \left|(\widetilde{L}_{\omega_k,+} f_k, f_k)_{L^2}\right| \le \lim_{k\to\infty} \|\widetilde{L}_{\omega_k,+} f_k\|_{L^2} = 0. \tag{2.32}$$

This is a contradiction. Thus, we have proved that (2.26) holds. □

LEMMA 2.0.7. *Assume $d \ge 3$ and $1 + \frac{4}{d} < p < 2^* - 1$. Then, we have*

$$\lim_{\omega\downarrow 0} \|\omega T_\omega \partial_\omega \Phi_\omega - \Lambda U\|_{H^1} = 0. \tag{2.33}$$

PROOF OF LEMMA 2.0.7. Let $\omega_0 > 0$ and $C > 0$ be constants given in Lemma 2.0.6. Furthermore, let $\{\omega_n\}$ be a sequence in $(0, \omega_0)$ with $\lim_{n\to\infty}\omega_n = 0$, and set

$$\Lambda\Phi_{\omega_n} := \partial_\omega \Phi_\omega|_{\omega=\omega_n}. \tag{2.34}$$

Then, we find from Proposition 2.0.6, (2.25) and Proposition 2.0.3 that there exists a number N such that for any $n \ge N$,

$$C\|\omega_n T_{\omega_n}\Lambda\Phi_{\omega_n}\|_{H^1} \le \left\|\widetilde{L}_{\omega_n,+}(\omega_n T_{\omega_n}\Lambda\Phi_{\omega_n})\right\|_{L^2} = \|T_{\omega_n}\Phi_{\omega_n}\|_{H^1} \le 2\|U\|_{H^1}. \tag{2.35}$$

Thus, $\{\omega_n T_{\omega_n}\Lambda\Phi_{\omega_n}\}$ is bounded in $H^1(\mathbb{R}^d)$ and therefore we can take $g_\infty \in H^1(\mathbb{R}^d)$ such that, passing to some subsequence,

$$\lim_{n\to\infty} \omega_n T_{\omega_n}\Lambda\Phi_{\omega_n} = g_\infty \qquad \text{weakly in } H^1(\mathbb{R}^d). \tag{2.36}$$

This together with Proposition 2.0.3 also shows that for any $\phi \in C_0^\infty(\mathbb{R}^d)$,

$$\lim_{n\to\infty} \langle \widetilde{L}_{\omega_n,+}(\omega_n T_{\omega_n}\Lambda\Phi_{\omega_n}), \phi\rangle_{H^{-1},H^1} = \langle L_+^\dagger g_\infty, \phi\rangle_{H^{-1},H^1}. \tag{2.37}$$

On the other hand, we see from (2.25) and Proposition 2.0.3 that

$$\lim_{n\to\infty} \widetilde{L}_{\omega_n,+}(\omega_n T_{\omega_n}\Lambda\Phi_{\omega_n}) = -U \qquad \text{strongly in } H^1(\mathbb{R}^d). \tag{2.38}$$

Putting (2.37) and (2.38) together, we find that $L_{+}^{\dagger} g_{\infty} = -U$. Furthermore, since $(\operatorname{Ker} L_{+}^{\dagger})\big|_{H^1_{rad}} = \{0\}$ (see Lemma 2.0.5), this identity together with (2.23) shows that $g_{\infty} = \Lambda U$. It remains to show that

$$\lim_{n\to\infty} \|\omega_n T_{\omega_n} \Lambda \Phi_{\omega_n}\|_{H^1} = \|\Lambda U\|_{H^1}. \tag{2.39}$$

We see from (2.25), Proposition 2.0.2, Proposition 2.0.3, (2.36) with $g_{\infty} = \Lambda U$ and (2.23) that

$$\begin{aligned}
&\lim_{n\to\infty} \|\omega_n T_{\omega_n} \Lambda \Phi_{\omega_n}\|_{H^1}^2 \\
&= -\lim_{n\to\infty} (T_{\omega_n}\Phi_{\omega_n}, \omega_n T_{\omega_n} \Lambda\Phi_{\omega_n})_{L^2_{real}} \\
&\quad + \lim_{n\to\infty} p((T_{\omega_n}\Phi_{\omega_n})^{p-1} \omega_n T_{\omega_n}\Lambda\Phi_{\omega_n}, \omega_n T_{\omega_n}\Lambda\Phi_{\omega_n})_{L^2_{real}} \\
&\quad + \lim_{n\to\infty} \omega_n^{\frac{2^*-(p+1)}{p-1}} (2^*-1)((T_{\omega_n}\Phi_{\omega_n})^{\frac{4}{d}} \omega_n T_{\omega_n}\Lambda\Phi_{\omega_n}, \omega_n T_{\omega_n}\Lambda\Phi_{\omega_n})_{L^2_{real}} \\
&= -(U, \Lambda U)_{L^2_{real}} + p(U^{p-1}\Lambda U, \Lambda U)_{L^2_{real}} \\
&= (L_{+}^{\dagger}\Lambda U, \Lambda U)_{L^2_{real}} + p(U^{p-1}\Lambda U, \Lambda U)_{L^2_{real}} = \|\Lambda U\|_{H^1}^2.
\end{aligned} \tag{2.40}$$

Thus, we have completed the proof. □

Now, we are in a position to prove the claim (iii) in Proposition 2.0.4.

Proof of (iii) in Proposition 2.0.4. We see from Proposition 2.0.3, (2.0.7) and (2.22) that

(2.41)

$$\begin{aligned}
\lim_{\omega\to 0} \omega^{s_p+1} \frac{d}{d\omega}\mathcal{M}(\Phi_{\omega}) &= \lim_{\omega\to 0} \omega^{s_p+1} (\Phi_{\omega}, \Phi_{\omega}')_{L^2_{real}} = \lim_{\omega\to 0} (T_{\omega}\Phi_{\omega}, \omega T_{\omega}\partial_{\omega}\Phi_{\omega})_{L^2_{real}} \\
&= (U, \Lambda U)_{L^2_{real}} = -\frac{s_p}{2}\|U\|_{L^2}^2 < 0,
\end{aligned}$$

which gives us the desired result. □

CHAPTER 3

Proof of Theorem 1.0.1

In order to prove Theorem 1.0.1, for a given $\varepsilon \geq 0$, we consider the set

$$A_{\omega}^{\varepsilon} := \{u \in H^1(\mathbb{R}^d) \colon \mathcal{S}_{\omega}(u) < m_{\omega} + \varepsilon,\ \mathcal{M}(u) = \mathcal{M}(\Phi_{\omega})\}. \tag{3.1}$$

Then, a key fact to prove Theorem 1.0.1 is the following:

THEOREM 3.0.1. *Assume either $d = 3$ and $3 \leq p < 5$, or $d \geq 4$ and $1 + \frac{4}{d-1} < p < 2^* - 1$. Then, there exists $\omega_* > 0$ with the following property: for any $\omega \in (0, \omega_*)$, there exists a positive constant $\varepsilon(\omega)$ such that all radial solutions starting from $A_{\omega}^{\varepsilon(\omega)}$ exhibit one of the following scenarios:*
(i) *Scattering both forward and backward in time;*
(ii) *Finite time blowup both forward and backward in time;*
(iii) *Scattering forward in time, and finite time blowup backward in time;*
(iv) *Finite time blowup forward in time, and scattering backward in time;*
(v) *Trapped by $\mathcal{O}(\Phi_{\omega})$ forward in time, and scattering backward in time;*
(vi) *Scattering forward in time, and trapped by $\mathcal{O}(\Phi_{\omega})$ backward in time;*
(vii) *Trapped by $\mathcal{O}(\Phi_{\omega})$ forward in time, and finite time blowup backward in time;*
(viii) *Finite time blowup forward in time, and trapped by $\mathcal{O}(\Phi_{\omega})$ backward in time;*
(ix) *Trapped by $\mathcal{O}(\Phi_{\omega})$ both forward and backward in time.*

We give a proof of this theorem in Section 8.

In order to prove Theorem 1.0.1, we also use the following fact:

LEMMA 3.0.1. *Assume $d \geq 3$ and $1 + \frac{4}{d} < p < 2^* - 1$. Let ω_1 be the frequency given by Proposition 2.0.4. Then, we have the following:*
(i) *m_{ω} is differentiable on $(0, \omega_1)$, and*

$$\frac{dm_{\omega}}{d\omega} = \mathcal{M}(\Phi_{\omega}) \tag{3.2}$$

for all $\omega \in (0, \omega_1)$. In particular, m_{ω} is strictly increasing on $(0, \omega_1)$.
(ii) *$\frac{m_{\omega}}{\omega}$ is differentiable and strictly decreasing on $(0, \omega_1)$.*
(iii) *Let $0 < \alpha < \beta < \omega_1$. Then, we have that*

$$\mathcal{M}(\Phi_{\beta}) < \frac{m_{\beta} - m_{\alpha}}{\beta - \alpha} < \mathcal{M}(\Phi_{\alpha}). \tag{3.3}$$

PROOF OF LEMMA 3.0.1. We shall prove the claim (i). Since $\mathcal{S}'(\Phi_{\omega}) = 0$ and Φ_{ω} is differentiable with respect to ω on $(0, \omega_1)$ (see (ii) of Proposition 2.0.4), we see that for any $\omega \in (0, \omega_1)$,

$$\frac{dm_{\omega}}{d\omega} = \frac{d}{d\omega}\mathcal{S}_{\omega}(\Phi_{\omega}) = \mathcal{M}(\Phi_{\omega}) + \mathcal{S}_{\omega}'(\Phi_{\omega})\partial_{\omega}\Phi_{\omega} = \mathcal{M}(\Phi_{\omega}). \tag{3.4}$$

Thus, we have proved the first claim.

Next, we shall prove the second claim (ii). The differentiability of $\frac{m_\omega}{\omega}$ follows from the first claim. Furthermore, we conclude from (ii) of Proposition 2.0.4 that for any $\omega \in (0, \omega_1)$,

$$\frac{d}{d\omega}\Big(\frac{m_\omega}{\omega}\Big) = \frac{d}{d\omega}\Big(\frac{\mathcal{S}_\omega(\Phi_\omega)}{\omega}\Big) = \frac{\mathcal{S}'_\omega(\Phi_\omega)\partial_\omega \Phi_\omega \omega - \mathcal{S}_\omega(\Phi_\omega)}{\omega^2} = -\frac{\mathcal{S}_\omega(\Phi_\omega)}{\omega^2} < 0. \tag{3.5}$$

Thus, we find that $\frac{m_\omega}{\omega}$ is strictly decreasing on $(0, \omega_1)$.

Finally, we shall prove the last claim (iii). It follows from the mean value theorem and (3.2) that for any $0 < \alpha < \beta < \omega_1$, there exists $\theta \in (0,1)$ such that

$$\frac{m_\beta - m_\alpha}{\beta - \alpha} = \mathcal{M}(\Phi_{\alpha+\theta(\beta-\alpha)}). \tag{3.6}$$

On the other hand we see from (iii) of Proposition 2.0.4 that for any $\theta \in (0,1)$,

$$\mathcal{M}(\Phi_\beta) < \mathcal{M}(\Phi_{\alpha+\theta(\beta-\alpha)}) < \mathcal{M}(\Phi_\alpha). \tag{3.7}$$

Putting (3.6) and (3.7) together, we obtain the desired result (3.3). □

We see from (2.9) with $q = 1$ and (iii) of Proposition 2.0.4 that there is a strictly decreasing function $\alpha\colon (\mathcal{M}(\Phi_{\omega_1}), \infty) \to (0, \omega_1)$ such that $M = \mathcal{M}(\Phi_{\alpha(M)})$ for any $M \in (\mathcal{M}(\Phi_{\omega_1}), \infty)$; α is the inverse of the function $\omega \mapsto \mathcal{M}(\Phi_\omega)$. Let ω_* be the frequency given by Theorem 3.0.1. Then, for each $\omega \in (0, \omega_*)$, we define a positive function $\varepsilon_\omega\colon [0,\infty) \to (0,\infty)$ to be that for any $M \geq 0$,

$$\varepsilon_\omega(M) := \begin{cases} m_{\omega_*} - m_\omega - (\omega_* - \omega)\mathcal{M}(\Phi_{\omega_*}) & \text{if } M \leq \mathcal{M}(\Phi_{\omega_*}), \\ \varepsilon(\alpha(M)) + m_{\alpha(M)} - m_\omega - (\alpha(M) - \omega)M & \text{if } \begin{array}{l}\mathcal{M}(\Phi_{\omega_*}) < M \\ \quad < \mathcal{M}(\Phi_\omega)\end{array}, \\ \varepsilon(\omega) & \text{if } M = \mathcal{M}(\Phi_\omega), \\ \varepsilon(\alpha(M)) + (\omega - \alpha(M))M - (m_\omega - m_{\alpha(M)}) & \text{if } M > \mathcal{M}(\Phi_\omega), \end{cases} \tag{3.8}$$

where $\varepsilon(\alpha(M))$ and $\varepsilon(\omega)$ are the positive constants given by Theorem 3.0.1.

REMARK 3.0.1. *It follows from* (iii) *of Lemma* 3.0.1 *that $\varepsilon_\omega(M)$ is positive for all $M \geq 0$. Furthermore, the continuity of $\varepsilon(\omega)$ with respect to ω implies that*

$$\inf_{M\geq 0} \varepsilon_\omega(M) > 0. \tag{3.9}$$

Now, we are in a position to prove Theorem 1.0.1.

PROOF OF THEOREM 1.0.1. Let ω_* be the frequency found in Theorem 3.0.1, and let $\omega \in (0, \omega_*)$. Furthermore, let ε_ω be the positive function define by (3.8), and let $\widetilde{PW}_\omega$ be the set defined by (1.16).

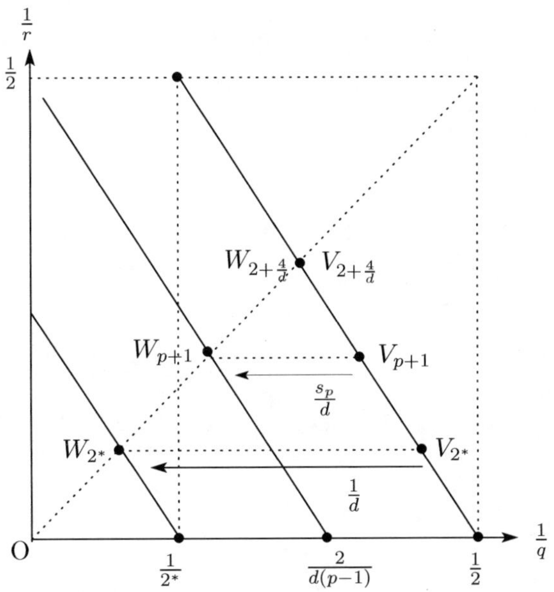

FIGURE 1. How to extend the potential well PW_ω

We consider a solution ψ to (NLS) starting from $\widetilde{PW}_\omega$. If $\mathcal{M}(\psi) \le \mathcal{M}(\Phi_{\omega_*})$, then we can verify that $\psi \in PW_{\omega_*}$ (see Figure 2 above). Hence, it follows from Theorem 1.0.3 that ψ exhibits the scenario (i) or (ii) in Theorem 1.0.1. On the other hand, if $\mathcal{M}(\Phi_{\omega_*}) < \mathcal{M}(\psi) < \mathcal{M}(\Phi_\omega)$, then we can verify that $\psi \in A_\alpha^{\varepsilon(\alpha)}$, where $\alpha := \alpha(\mathcal{M}(\psi))$. Similarly, if $\mathcal{M}(\psi) \ge \mathcal{M}(\Phi_\omega)$, then $\psi \in A_\alpha^{\varepsilon(\alpha)}$. Hence, it follows from Theorem 3.0.1 that ψ exhibits one of the nine scenarios in Theorem 1.0.1. □

CHAPTER 4

Decomposition around ground state

In this section, for a positive radial ground state Φ_ω to (1.7) and a solution ψ to (NLS) with the maximal existence-interval $I_{\max}$, we consider the decomposition of the form

$$\psi(x,t) = e^{i\theta(t)}\big(\Phi_\omega(x) + \eta(x,t)\big), \tag{4.1}$$

where $\theta(t)$ is some function of $t \in I_{\max}$ to be chosen later, and η is the remainder. In this decomposition, we do not assume the orthogonality between Φ_ω and $\eta(t)$ in L^2_{real} (see (4.57) below).

4.0.1. Linearized operator. The decomposition (4.1) leads us to the linearized operator $\mathcal{L}_\omega$ around Φ_ω which is defined by

(4.2)

$$\begin{aligned}\mathcal{L}_\omega u &:= \omega u - \Delta u - \frac{p+1}{2}\Phi_\omega^{p-1}u - \frac{p-1}{2}\Phi_\omega^{p-1}\overline{u} - \frac{2^*}{2}\Phi_\omega^{\frac{4}{d-2}}u - \frac{2^*-2}{2}\Phi_\omega^{\frac{4}{d-2}}\overline{u}\\ &= \big(\omega - \Delta - p\Phi_\omega^{p-1} - (2^*-1)\Phi_\omega^{\frac{4}{d-2}}\big)\Re[u] + i\big(\omega - \Delta - \Phi_\omega^{p-1} - \Phi_\omega^{\frac{4}{d-2}}\big)\Im[u].\end{aligned}$$

The operator $\mathcal{L}_\omega$ is self-adjoint in $L^2_{real}(\mathbb{R}^d)$, and for any $u, v \in H^1(\mathbb{R}^d)$,

$$\big[\mathcal{S}_\omega''(\Phi_\omega)u\big]v = \langle \mathcal{L}_\omega u, v\rangle_{H^{-1},H^1}. \tag{4.3}$$

It is convenient to introduce the operators $L_{\omega,+}$ and $L_{\omega,-}$:

$$L_{\omega,+} := \omega - \Delta - p\Phi_\omega^{p-1} - (2^*-1)\Phi_\omega^{\frac{4}{d-2}}, \tag{4.4}$$

$$L_{\omega,-} := \omega - \Delta - \Phi_\omega^{p-1} - \Phi_\omega^{\frac{4}{d-2}}. \tag{4.5}$$

Then, we have

$$\mathcal{L}_\omega u = L_{\omega,+}\Re[u] + iL_{\omega,-}\Im[u]. \tag{4.6}$$

Moreover, since Φ_ω is a solution to (1.7), we can verify that

$$\mathcal{L}_\omega\Phi_\omega = L_{\omega,+}\Phi_\omega = -(p-1)\Phi_\omega^p - (2^*-2)\Phi_\omega^{2^*-1}, \tag{4.7}$$

$$\mathcal{L}_\omega(i\Phi_\omega) = L_{\omega,-}\Phi_\omega = 0, \tag{4.8}$$

$$\mathcal{L}_\omega\partial_\omega\Phi_\omega = L_{\omega,+}\partial_\omega\Phi_\omega = -\Phi_\omega. \tag{4.9}$$

Inserting the decomposition $\psi(t) = e^{i\theta(t)}(\Phi_\omega + \eta(t))$ into the equation (NLS), we obtain the equation for η:

$$\frac{\partial\eta}{\partial t}(t) = -i\mathcal{L}_\omega\eta(t) - i\Big\{\frac{d\theta}{dt}(t) - \omega\Big\}(\Phi_\omega + \eta(t)) + iN_\omega(\eta(t)), \tag{4.10}$$

where $N_\omega(\eta)$ denotes the higher order term of η, i.e.,

$$\begin{aligned} N_\omega(\eta) &:= |\Phi_\omega+\eta|^{p-1}(\Phi_\omega+\eta) - |\Phi_\omega|^{p-1}\Phi_\omega - \frac{p+1}{2}\Phi_\omega^{p-1}\eta - \frac{p-1}{2}\Phi_\omega^{p-1}\overline{\eta} \\ &\quad + |\Phi_\omega+\eta|^{\frac{4}{d-2}}(\Phi_\omega+\eta) - |\Phi_\omega|^{\frac{4}{d-2}}\Phi_\omega - \frac{2^*}{2}\Phi_\omega^{\frac{4}{d-2}}\eta - \frac{2^*-2}{2}\Phi_\omega^{\frac{4}{d-2}}\overline{\eta} \\ &= O\big(|\eta|^{\min\{2,p\}}\big). \end{aligned} \tag{4.11}$$

We see from (4.10) that the operator $-i\mathcal{L}_\omega$ relates to the behavior of the remainder $\eta(t)$. Unfortunately, $-i\mathcal{L}_\omega$ is not symmetric in $L^2_{real}(\mathbb{R}^d)$, and therefore we do not have the orthogonality of eigenfunctions in this space. Thus, we need to work in the symplectic space $(L^2(\mathbb{R}^d),\Omega)$ instead of $L^2_{real}(\mathbb{R}^2)$, where Ω is the symplectic form defined by

$$\Omega(f,g) := \Im\int_{\mathbb{R}^d} f(x)\overline{g(x)}\,dx = (f, ig)_{L^2_{real}}. \tag{4.12}$$

Proposition 4.0.1. *Assume $d\ge 3$ and $1+\frac{4}{d}<p<2^*-1$. Let ω_1 be the frequency given by Proposition 2.0.4. Then, there exists $\omega_2\in(0,\omega_1)$ such that for any $\omega\in(0,\omega_2)$, $-i\mathcal{L}_\omega$ has a positive eigenvalue μ, as an operator in $L^2_{real}(\mathbb{R}^d)$. Furthermore, the eigenvalue μ satisfies that*

$$-\mu^2 = \inf\left\{ \frac{\langle L_{\omega,+}u,u\rangle_{H^{-1},H^1}}{((L_{\omega,-})^{-1}u,u)_{L^2_{real}}} : u\in H^1(\mathbb{R}^d),\ (u,\Phi_\omega)_{L^2_{real}}=0 \right\}. \tag{4.13}$$

Proof of Proposition 4.0.1. We prove the proposition following the exposition in Section 2 of [**9**]. What we need to prove is that there exist a function $u\in H^1(\mathbb{R}^d)$ and $\nu<0$ such that

$$L_{\omega,-}L_{\omega,+}u = \nu u. \tag{4.14}$$

Indeed, putting $f_1=-\Re[u]$ and $f_2=-\sqrt{-\nu}(L_{\omega,-})^{-1}\Re[u]$, we see from (4.6) and (4.14) that

(4.15)

$$\begin{aligned} -i\mathcal{L}_\omega(f_1+if_2) &= -iL_{\omega,+}f_1 + L_{\omega,-}f_2 \\ &= i(L_{\omega,-})^{-1}\nu\Re[u] - L_{\omega,-}\sqrt{-\nu}(L_{\omega,-})^{-1}\Re[u] = \sqrt{-\nu}(f_1+if_2). \end{aligned}$$

Furthermore, we can verify that the problem (4.14) is equivalent to that there exist $u\in H^1(\mathbb{R}^d)$, $\nu<0$ and $\alpha\in\mathbb{R}$ such that

$$\begin{cases} L_{\omega,+}u &= \nu(L_{\omega,-})^{-1}u + \alpha\Phi_\omega, \\ (u,\Phi_\omega)_{L^2_{real}} &= 0. \end{cases} \tag{4.16}$$

Note here that $\operatorname{Ker} L_{\omega,-} = \operatorname{span}\{\Phi_\omega\}$ (see Lemma B.0.1) and $-\Phi_\omega = L_{\omega,+}\partial_\omega\Phi_\omega$ (see (4.9)). The problem (4.16) leads us to the following minimizing problem

$$\nu_\omega := \inf\left\{ \frac{\langle L_{\omega,+}u,u\rangle_{H^{-1},H^1}}{((L_{\omega,-})^{-1}u,u)_{L^2_{real}}} : u\in H^1(\mathbb{R}^d),\ (u,\Phi_\omega)_{L^2_{real}}=0 \right\}. \tag{4.17}$$

We can verify that any minimizer of (4.17) satisfies the equation (4.16) with $\nu=\nu_\omega$ for some $\alpha\in\mathbb{R}$. Thus, it suffices to show that $-\infty<\nu_\omega<0$ and the existence of a minimizer.

First, we shall show that $-\infty < \nu_\omega$. We take any $u \in H^1(\mathbb{R}^d)$ with $(u, \Phi_\omega)_{L^2_{real}} = 0$. Then, we see from Proposition 2.0.2 that

$$\begin{aligned}
&\frac{\langle L_{\omega,+}u, u\rangle_{H^{-1},H^1}}{((L_{\omega,-})^{-1}u, u)_{L^2_{real}}} \\
&= \frac{\langle L_{\omega,-}u, u\rangle_{H^{-1},H^1}}{\|(L_{\omega,-})^{-\frac{1}{2}}u\|_{L^2}^2} - \frac{\langle \{(p-1)\Phi_\omega^{p-1} + (2^*-2)\Phi_\omega^{\frac{4}{d-2}}\}u, u\rangle_{H^{-1},H^1}}{\|(L_{\omega,-})^{-\frac{1}{2}}u\|_{L^2}^2} \\
&\ge \frac{\|(L_{\omega,-})^{\frac{1}{2}}u\|_{H^1}^2}{\|(L_{\omega,-})^{-\frac{1}{2}}u\|_{L^2}^2} - C(\omega)\frac{\|u\|_{L^2}^2}{\|(L_{\omega,-})^{-\frac{1}{2}}u\|_{L^2}^2},
\end{aligned} \tag{4.18}$$

where $C(\omega)$ is some positive constant depending on ω. Moreover, it is easy to see that

$$\begin{aligned}
\|u\|_{L^2}^2 &= ((L_{\omega,-})^{\frac{1}{2}}u, (L_{\omega,-})^{-\frac{1}{2}}u)_{L^2_{real}} \le \|(L_{\omega,-})^{\frac{1}{2}}u\|_{L^2}\|(L_{\omega,-})^{-\frac{1}{2}}u\|_{L^2} \\
&\le \frac{1}{2C(\omega)}\|(L_{\omega,-})^{\frac{1}{2}}u\|_{L^2}^2 + \frac{C(\omega)}{2}\|(L_{\omega,-})^{-\frac{1}{2}}u\|_{L^2}^2.
\end{aligned} \tag{4.19}$$

Putting the estimates (4.18) and (4.19) together, we find that $\nu_\omega > -C(\omega)^2 > -\infty$.

Next, we shall show that ν_ω is negative for any sufficiently small $\omega > 0$. To this end, we introduce projections Π_ω and Π:

$$\Pi_\omega u := u - \frac{(u, \Phi_\omega)_{L^2_{real}}}{\|\Phi_\omega\|_{L^2}^2}\Phi_\omega, \quad \Pi u := u - \frac{(u, U)_{L^2_{real}}}{\|U\|_{L^2}^2}U. \tag{4.20}$$

Then, we see from substitution of variables, Proposition 2.0.3 and Lemma 2.0.7 that

$$\begin{aligned}
\omega^{-2}\nu_\omega &= \omega^{-2}\inf\left\{\frac{\langle L_{\omega,+}u, u\rangle_{H^{-1},H^1}}{((L_{\omega,-})^{-1}u, u)_{L^2_{real}}} : u \in H^1(\mathbb{R}^d),\ (u, \Phi_\omega)_{L^2_{real}} = 0\right\} \\
&\le \omega^{-2}\frac{\langle L_{\omega,+}\Pi_\omega\partial_\omega\Phi_\omega, \Pi_\omega\partial_\omega\Phi_\omega\rangle_{H^{-1},H^1}}{((L_{\omega,-})^{-1}\Pi_\omega\partial_\omega\Phi_\omega, \Pi_\omega\partial_\omega\Phi_\omega)_{L^2_{real}}} \\
&= \frac{\langle \widetilde{L}_{\omega,+}(\omega T_\omega\Pi_\omega\partial_\omega\Phi_\omega), \omega T_\omega\Pi_\omega\partial_\omega\Phi_\omega\rangle_{H^{-1},H^1}}{((\widetilde{L}_{\omega,-})^{-1}(\omega T_\omega\Pi_\omega\partial_\omega\Phi_\omega), \omega T_\omega\Pi_\omega\partial_\omega\Phi_\omega)_{L^2_{real}}} \\
&\to \frac{\langle L_+^\dagger\Pi\Lambda U, \Pi\Lambda U\rangle_{H^{-1},H^1}}{((L_-^\dagger)^{-1}\Pi\Lambda U, \Pi\Lambda U)_{L^2_{real}}} \quad \text{as } \omega \downarrow 0,
\end{aligned} \tag{4.21}$$

where $\widetilde{L}_{\omega,+}$, $\widetilde{L}_{\omega,-}$, $L_+^\dagger$ and $L_-^\dagger$ are operators defined by (1.59), (1.60), (1.55) and (1.56), respectively. Thus, it suffices to show that

$$\frac{\langle L_+^\dagger\Pi\Lambda U, \Pi\Lambda U\rangle_{H^{-1},H^1}}{\langle (L^\dagger)_-^{-1}\Pi\Lambda U, \Pi\Lambda U)_{L^2_{real}}} < 0. \tag{4.22}$$

Note here that we see from (2.22) and integration by parts that

$$\Pi\Lambda U = \Lambda U + \frac{s_p}{2}U. \tag{4.23}$$

Moreover, it follows from (2.22) and (2.23) that

$$L_{+}^{\dagger}\Pi\Lambda U = L_{+}^{\dagger}\Lambda U - \frac{(\Lambda U, U)_{L^2_{real}}}{\|U\|_{L^2}^2} L_{+}^{\dagger} U = -U - \frac{(p-1)s_p}{2} U^p. \tag{4.24}$$

Using (4.23), (4.24), (2.22) and integration by parts, we obtain that

(4.25)

$$\begin{aligned}
\langle L_{+}^{\dagger}\Pi\Lambda U, \Pi\Lambda U\rangle_{H^{-1},H^1} &= \langle -U - \frac{(p-1)s_p}{2} U^p, \Lambda U + \frac{s_p}{2} U\rangle_{H^{-1},H^1} \\
&= -(U, \Lambda U)_{L^2_{real}} - \frac{(p-1)s_p}{2} (U^p, \Lambda U)_{L^2_{real}} - \frac{s_p}{2}\|U\|_{L^2}^2 - \frac{(p-1)s_p^2}{4}\|U\|_{L^{p+1}}^{p+1} \\
&= \frac{d(p-1)s_p}{4}\Big(\frac{1}{2} - \frac{1}{p+1}\Big)\|U\|_{L^p}^p < 0.
\end{aligned}$$

Since Lemma B.0.1 implies that $(\Pi\Lambda U, (L_{-}^{\dagger})^{-1}\Pi\Lambda U)_{L^2_{real}} > 0$, we find from (4.25) that (4.22) holds.

Finally, we shall show the existence of a minimizer u for ν_ω. We can take a sequence $\{u_n\}$ in $H^1(\mathbb{R}^d)$ such that

$$(u_n, \Phi_\omega)_{L^2_{real}} \equiv 0, \tag{4.26}$$

$$((L_{\omega,-})^{-1} u_n, u_n)_{L^2_{real}} \equiv 1, \tag{4.27}$$

$$\lim_{n\to\infty} \langle L_{\omega,+} u_n, u_n\rangle_{H^{-1},H^1} = \nu_\omega. \tag{4.28}$$

Note here that Lemma B.0.1 shows that the square root of $L_{\omega,-}$ is well-defined. We see from (4.28), Proposition 2.0.2, the Cauchy-Schwarz inequality, (B.1) and (4.27) that for any sufficiently large number n,

$$\begin{aligned}
\|u_n\|_{H^1}^2 &\le (1+\omega^{-1})\langle L_{\omega,+} u_n, u_n\rangle_{H^{-1},H^1} \\
&\qquad + \int_{\mathbb{R}^d} \{p(\Phi_\omega(x))^{p-1} + (2^*-1)(\Phi_\omega(x))^{\frac{4}{d-2}}\}|u_n(x)|^2\,dx \\
&\lesssim (1+\omega^{-1})\nu_\omega + C_1(\omega)((L_{\omega,-})^{\frac{1}{2}} u_n, (L_{\omega,-})^{-\frac{1}{2}} u_n)_{L^2_{real}} \\
&\le (1+\omega^{-1})\nu_\omega + C_1(\omega)\|(L_{\omega,-})^{\frac{1}{2}} u_n\|_{L^2}\|(L_{\omega,-})^{-\frac{1}{2}} u_n\|_{L^2} \\
&= (1+\omega^{-1})\nu_\omega + C_1(\omega)(L_{\omega,-} u_n, u_n)_{L^2_{real}}^{\frac{1}{2}}((L_{\omega,-})^{-1} u_n, u_n)_{L^2_{real}}^{\frac{1}{2}} \\
&\le (1+\omega^{-1})\nu_\omega + C_2(\omega)\|u_n\|_{H^1},
\end{aligned} \tag{4.29}$$

where $C_1(\omega)$ and $C_2(\omega)$ are some constants depending only on d, p and ω. This implies that $\{u_n\}$ is bounded in $H^1(\mathbb{R}^d)$ and therefore we can extract a subsequence of $\{u_n\}$ (still denoted by the same symbol) and a function $u_\infty \in H^1(\mathbb{R}^d)$ such that

$$\lim_{n\to\infty} u_n = u_\infty \qquad \text{weakly in } H^1(\mathbb{R}^d), \tag{4.30}$$

$$(u_\infty, \Phi_\omega)_{L^2_{real}} = 0, \tag{4.31}$$

$$\lim_{n\to\infty} \int_{\mathbb{R}^d} \Phi_\omega(x)^q |u_n(x)|^2 \, dx = \int_{\mathbb{R}^d} \Phi_\omega(x)^q |u_\infty(x)|^2 \, dx, \tag{4.32}$$

where q indicates $p-1$ or $\frac{4}{d-2}$: Here, (4.31) follows from (4.26) and (4.30), and (4.32) follows from (4.30) and Propositionre 2.0.2. Furthermore, we can verify that

$$\langle L_{\omega,+}u_\infty, u_\infty\rangle_{H^{-1},H^1} \le \liminf_{n\to\infty} \langle L_{\omega,+}u_n, u_n\rangle_{H^{-1},H^1} = \nu_\omega < 0. \tag{4.33}$$

In particular, the limit u_∞ is non-trivial. We also see from Lemma B.0.1, (4.8), (4.26) and (4.27) that for any number n,

$$\|(L_{\omega,-})^{-1}u_n\|_{H^1}^2 \lesssim (u_n, (L_{\omega,-})^{-1}u_n)_{L^2_{real}} \le 1, \tag{4.34}$$

where the implicit constant depends on ω. Hence, there exists a subsequence of $\{u_n\}$ (still denoted by the same symbol) and a function $v_\infty \in H^1(\mathbb{R}^d)$ such that

$$\lim_{n\to\infty} (L_{\omega,-})^{-1}u_n = v_\infty \qquad \text{weakly in } H^1(\mathbb{R}^d). \tag{4.35}$$

Furthermore, we see from (4.8) that for any test function $\phi \in C_c^\infty(\mathbb{R}^d)$,

$$\begin{aligned} \langle L_{\omega,-}v_\infty, \phi\rangle_{H^{-1},H^1} &= \langle L_{\omega,-}\phi, v_\infty\rangle_{H^{-1},H^1} \\ &= \lim_{n\to\infty} ((L_{\omega,-})^{-1}u_n, L_{\omega,-}\phi)_{L^2_{real}} = (u_\infty, \phi)_{L^2_{real}}. \end{aligned} \tag{4.36}$$

This together with (4.31) shows that

$$v_\infty = (L_{\omega,-})^{-1}u_\infty. \tag{4.37}$$

We also see from (4.37) and Lemma B.0.1 that

$$((L_{\omega,-})^{-1}u_\infty, u_\infty)_{L^2_{real}} = (v_\infty, L_{\omega,-}v_\infty)_{L^2_{real}} > 0. \tag{4.38}$$

Furthermore, it follows from (4.35), (4.37), (4.26) and (4.27) that

$$((L_{\omega,-})^{-1}u_\infty, u_\infty)_{L^2_{real}} \le \liminf_{n\to\infty} ((L_{\omega,-})^{-1}u_n, u_n)_{L^2_{real}} = 1, \tag{4.39}$$

so that we can take $\rho_1 \ge 1$ such that $((L_{\omega,-})^{-1}\rho_1 u_\infty, \rho_1 u_\infty)_{L^2_{real}} = 1$. Suppose here that $\rho_1 > 1$. Then, we see from (4.31) and (4.33) that

$$\nu_\omega \le \langle L_{\omega,+}\rho_1 u_\infty, \rho_1 u_\infty\rangle_{H^{-1},H^1} < \langle L_{\omega,+}u_\infty, u_\infty\rangle_{H^{-1},H^1} \le \nu_\omega. \tag{4.40}$$

This is a contradiction. Thus, we have shown that

$$((L_{\omega,-})^{-1}u_\infty, u_\infty)_{L^2_{real}} = 1. \tag{4.41}$$

Then, the same argument as (4.40) shows that

$$\nu_\omega = \langle L_{\omega,+}u_\infty, u_\infty\rangle_{H^{-1},H^1}. \tag{4.42}$$

This together with (4.31) and (4.41) shows that u_∞ is a minimizer for the problem (4.17). □

4.0.2. Symplectic decomposition. In this subsection, we assume that $d \geq 3$, $1+\frac{4}{d} < p < 2^*-1$ and $0 < \omega < \omega_2$, where ω_2 is the frequency given in Proposition 4.0.1. Note that $-i\mathcal{L}_\omega$ has a positive eigenvalue $\mu > 0$ as an operator in $L^2_{real}(\mathbb{R}^d)$.

We will apply the "symplectic decomposition" corresponding to the discrete modes of $-i\mathcal{L}_\omega$ to the remainder η in (4.1) (see (4.48) below). Moreover, we determine the function $\theta(t)$ in (4.1).

Let $\mathfrak{U}_+$ be an eigenfunction corresponding to the positive eigenvalue μ, and put

$$\mathfrak{U}_- := \overline{\mathfrak{U}_+}. \tag{4.43}$$

Then, we have

$$-i\mathcal{L}_\omega\mathfrak{U}_- = \overline{i\mathcal{L}_\omega\mathfrak{U}_+} = -\overline{\mu\mathfrak{U}_+} = -\mu\mathfrak{U}_-. \tag{4.44}$$

Hence, $\mathfrak{U}_-$ is an eigenfunction of $-i\mathcal{L}_\omega$ corresponding to $-\mu$. We assume that $\mathfrak{U}_+$ and $\mathfrak{U}_-$ are normalized in the following sense:

$$\Omega(\mathfrak{U}_+, \mathfrak{U}_-) = 1, \quad \Omega(\mathfrak{U}_-, \mathfrak{U}_+) = -1. \tag{4.45}$$

It is obvious that

$$\Omega(\mathfrak{U}_+, \mathfrak{U}_+) = \Omega(\mathfrak{U}_-, \mathfrak{U}_-) = 0. \tag{4.46}$$

Furthermore, it follows from (4.8), (4.9) and $\mathcal{L}_\omega\mathfrak{U}_\pm = \pm i\mu\mathfrak{U}_\pm$ that

$$\Omega(i\Phi_\omega, \mathfrak{U}_\pm) = \Omega(\partial_\omega\Phi_\omega, \mathfrak{U}_\pm) = 0. \tag{4.47}$$

Now, we expand the remainder $\eta(t)$ in the decomposition (4.1) by the discrete modes of $-i\mathcal{L}_\omega$:

$$\eta(t) = \lambda_+(t)\mathfrak{U}_+ + \lambda_-(t)\mathfrak{U}_- + a(t)i\Phi_\omega + b(t)\partial_\omega\Phi_\omega + \gamma(t), \tag{4.48}$$

where

$$\Omega(\gamma(t), \mathfrak{U}_+) = \Omega(\gamma(t), \mathfrak{U}_-) = \Omega(\gamma(t), i\Phi_\omega) = \Omega(\gamma(t), \partial_\omega\Phi_\omega) = 0. \tag{4.49}$$

We see from (4.45), (4.46) and (4.47) that the coefficients are as follows:

$$\lambda_+(t) = \Omega(\eta(t), \mathfrak{U}_-), \qquad \lambda_-(t) = -\Omega(\eta(t), \mathfrak{U}_+), \tag{4.50}$$

$$a(t) = \frac{\Omega(\eta(t), \partial_\omega\Phi_\omega)}{(\Phi_\omega, \partial_\omega\Phi_\omega)_{L^2_{real}}}, \qquad b(t) = -\frac{\Omega(\eta(t), i\Phi_\omega)}{(\Phi_\omega, \partial_\omega\Phi_\omega)_{L^2_{real}}}. \tag{4.51}$$

Note here that the denominators in (4.51) are non-zero (see (iii) of Proposition 2.0.4). Moreover, it follows from Hölder's inequality that

$$\lambda_\pm^2(t) = \Omega(\eta(t), \mathfrak{U}_\mp)^2 \leq \|\mathfrak{U}_+\|_{L^2}^2 \|\eta(t)\|_{L^2}^2. \tag{4.52}$$

We require that the mode $i\Phi_\omega$ does not appear in the decomposition (4.48), that is, $a(t) \equiv 0$. To this end, we choose the function $\theta(t)$ in (4.1) so that

$$\Omega(e^{-i\theta(t)}\psi(t), \partial_\omega\Phi_\omega) \equiv 0. \tag{4.53}$$

Then, it follows from $\Omega(\Phi_\omega, \partial_\omega\Phi_\omega) = 0$ that $\Omega(\eta(t), \partial_\omega\Phi_\omega) \equiv 0$, and therefore $a(t) \equiv 0$. Furthermore, the choice of $\theta(t)$ has room of an integer multiple of π. Hence, in addition to (4.53), we can choose $\theta(t)$ so that

$$(e^{-i\theta(t)}\psi(t), \partial_\omega\Phi_\omega)_{L^2_{real}} < 0. \tag{4.54}$$

This choice of $\theta(t)$ plays an important role in the argument below (see (6.10)).

4.0.3. Modulation equations. We continue the discussion about the decompositions (4.1) and (4.48) for a solution ψ to (NLS). Throughout this subsection, as well as Section 4.0.2, we assume that $d \ge 3$, $1+\frac{4}{d} < p < 2^*-1$ and $0<\omega<\omega_2$. Furthermore, we assume that ψ satisfies (4.53), (4.54) and

$$\mathcal{M}(\psi) = \mathcal{M}(\Phi_\omega). \tag{4.55}$$

Since

$$\mathcal{M}(\psi) = \mathcal{M}(\Phi_\omega) + \mathcal{M}(\eta(t)) + \big(\Phi_\omega, \eta(t)\big)_{L^2_{real}}, \tag{4.56}$$

the condition (4.55) implies that for any $t \in I_{\max}$,

$$\big(\Phi_\omega, \eta(t)\big)_{L^2_{real}} = -\mathcal{M}(\eta(t)). \tag{4.57}$$

In particular, there is no orthogonality between Φ_ω and $\eta(t)$ in $L^2_{real}(\mathbb{R}^d)$.

Our aim here is to derive ordinary differential equations for θ, λ_+ and λ_- under the condition (4.55).

First, we shall derive an equation for θ. We see from (4.53) that

$$0 = \frac{d}{dt}\Omega(\Phi_\omega + \eta(t), \partial_\omega \Phi_\omega) = (\frac{d\eta}{dt}(t), i\partial_\omega\Phi_\omega)_{L^2_{real}}. \tag{4.58}$$

Putting (4.10) and (4.58) together, we obtain

$$\begin{aligned} 0 &= (-\mathcal{L}_\omega \eta(t) - \Big\{\frac{d\theta}{dt}(t) - \omega\Big\}(\Phi_\omega + \eta(t)) + N_\omega(\eta(t)),\, \partial_\omega\Phi_\omega)_{L^2_{real}} \\ &= -\big(\eta(t), \mathcal{L}_\omega\partial_\omega\Phi_\omega\big)_{L^2_{real}} - \Big\{\frac{d\theta}{dt}(t) - \omega\Big\}\big(\Phi_\omega + \eta(t), \partial_\omega\Phi_\omega\big)_{L^2_{real}} \\ &\quad + \big(N_\omega(\eta(t)), \partial_\omega\Phi_\omega\big)_{L^2_{real}}. \end{aligned} \tag{4.59}$$

Furthermore, using (4.9) and (4.57), we obtain the equation for θ:

$$\Big\{\frac{d\theta}{dt}(t) - \omega\Big\}\big(\Phi_\omega + \eta(t), \partial_\omega\Phi_\omega\big)_{L^2_{real}} = -\mathcal{M}(\eta(t)) + \big(N_\omega(\eta(t)), \partial_\omega\Phi_\omega\big)_{L^2_{real}}. \tag{4.60}$$

Next, we shall derive equations for λ_+ and λ_-. It follows from (4.10), (4.47), (4.50) and $\mathcal{L}_\omega \mathfrak{U}_\pm = \pm i\mu \mathfrak{U}_\pm$ that

$$\begin{aligned} \frac{d\lambda_\pm}{dt}(t) &= \pm\Omega(\frac{d\eta}{dt}(t), \mathfrak{U}_\mp) \\ &= \pm\Omega(-i\mathcal{L}_\omega\eta(t) - i\Big\{\frac{d\theta}{dt}(t) - \omega\Big\}\eta(t) + iN_\omega(\eta(t)),\, \mathfrak{U}_\mp) \\ &= \pm\big(\eta(t), i\mu\mathfrak{U}_\mp\big)_{L^2_{real}} \mp \big(\Big\{\frac{d\theta}{dt}(t) - \omega\Big\}\eta(t) - N_\omega(\eta(t)),\, \mathfrak{U}_\mp\big)_{L^2_{real}} \\ &= \pm\mu\lambda_\pm(t) \mp \big(\Big\{\frac{d\theta}{dt}(t) - \omega\Big\}\eta(t) - N_\omega(\eta(t)),\, \mathfrak{U}_\mp\big)_{L^2_{real}}. \end{aligned} \tag{4.61}$$

Thus, we have obtained equations for λ_+ and λ_-:

$$\frac{d\lambda_+}{dt}(t) = \mu\lambda_+(t) - \big(\Big\{\frac{d\theta}{dt}(t) - \omega\Big\}\eta(t) - N_\omega(\eta(t)),\, \mathfrak{U}_-\big)_{L^2_{real}}, \tag{4.62}$$

$$\frac{d\lambda_-}{dt}(t) = -\mu\lambda_-(t) + \big(\Big\{\frac{d\theta}{dt}(t) - \omega\Big\}\eta(t) - N_\omega(\eta(t)),\, \mathfrak{U}_+\big)_{L^2_{real}}. \tag{4.63}$$

Note here that it follows from $i\mathcal{L}_\omega \mathfrak{U}_\pm = \pm\mu\mathfrak{U}_\pm$, (4.62), (4.63) and (4.45) that

$$\begin{aligned}&\langle \mathcal{L}_\omega\big(\lambda_+(t)\mathfrak{U}_+ + \lambda_-(t)\mathfrak{U}_-\big), \frac{d\lambda_+}{dt}(t)\mathfrak{U}_+ + \frac{d\lambda_-}{dt}(t)\mathfrak{U}_-\rangle_{H^{-1},H^1}\\ &= -\mu\lambda_-(t)\big(\Big\{\frac{d\theta}{dt}(t)-\omega\Big\}\eta(t) - N_\omega(\eta(t)),\, \mathfrak{U}_-\big)_{L^2_{real}}\\ &\quad + \mu\lambda_+(t)\big(\Big\{\frac{d\theta}{dt}(t)-\omega\Big\}\eta(t) - N_\omega(\eta(t)),\, \mathfrak{U}_+\big)_{L^2_{real}}.\end{aligned} \tag{4.64}$$

4.0.4. Linearized energy norm. Our aim here it to introduce the "linearized energy norm" for the remainder η in the decomposition (4.1). Throughout this subsection, as well as Section 4.0.3, we assume that $d \ge 3$, $1+\frac{4}{d} < p < 2^*-1$, $0<\omega<\omega_2$ and ψ is a solution to (NLS) satisfying (4.53) through (4.55).

Put

$$\Gamma(t) := b(t)\partial_\omega\Phi_\omega + \gamma(t). \tag{4.65}$$

Then, since $a(t) \equiv 0$ (see (4.53)), the decomposition (4.48) is rewritten by

$$\eta(t) = \lambda_+(t)\mathfrak{U}_+ + \lambda_-(t)\mathfrak{U}_- + \Gamma(t). \tag{4.66}$$

We see from (4.47) and (4.49) that

$$\Omega(\Gamma(t),\mathfrak{U}_+) = \Omega(\Gamma(t),\mathfrak{U}_-) = \Omega\big(\Gamma(t),\partial_\omega\Phi_\omega\big) = 0. \tag{4.67}$$

As a consequence of these orthogonalities, we have the following relationship:

LEMMA 4.0.2. *The function* Γ *in the decomposition* (4.66) *satisfies*

$$\langle \mathcal{L}_\omega\Gamma(t),\Gamma(t)\rangle_{H^{-1},H^1} \sim \|\Gamma(t)\|_{H^1}^2 \tag{4.68}$$

for all $t \in I_{\max}$, *where the implicit constant depends on* ω *as well as* d *and* p.

PROOF OF LEMMA 4.0.2. Lemma B.0.5 together with (4.67) gives the desired result. □

Now, we see from (4.55), Taylor's expansion around Φ_ω, $\mathcal{S}_\omega'(\Phi_\omega)=0$, (4.3), $\mathcal{L}_\omega\mathfrak{U}_\pm = \pm i\mu\mathfrak{U}_\pm$ and (4.45) that

$$\begin{aligned}&\mathcal{H}(\psi) - \mathcal{H}(\Phi_\omega) = \mathcal{S}_\omega(\psi) - \mathcal{S}_\omega(\Phi_\omega)\\ &= \mathcal{S}_\omega(e^{-i\theta(t)}\psi) - \mathcal{S}_\omega(\Phi_\omega) = \mathcal{S}_\omega(\Phi_\omega+\eta(t)) - \mathcal{S}_\omega(\Phi_\omega)\\ &= \frac{1}{2}\langle \mathcal{L}_\omega\eta(t),\eta(t)\rangle_{H^{-1},H^1} + O(\|\eta(t)\|_{H^1}^{\min\{3,p+1\}})\\ &= \frac{1}{2}\lambda_+(t)\lambda_-(t)(-i\mu\mathfrak{U}_+,\mathfrak{U}_-)_{L^2_{real}} + \frac{1}{2}\lambda_+(t)\lambda_-(t)(i\mu\mathfrak{U}_-,\,\mathfrak{U}_+)_{L^2_{real}}\\ &\quad + \frac{1}{2}\langle \mathcal{L}_\omega\Gamma(t),\Gamma(t)\rangle_{H^{-1},H^1} + O(\|\eta(t)\|_{H^1}^{\min\{3,p+1\}})\\ &= -\mu\lambda_+(t)\lambda_-(t) + \frac{1}{2}\langle \mathcal{L}_\omega\Gamma(t),\Gamma(t)\rangle_{H^{-1},H^1} + O(\|\eta(t)\|_{H^1}^{\min\{3,p+1\}}).\end{aligned} \tag{4.69}$$

We define the linearized energy norm $\|\eta(t)\|_E$ by

$$\|\eta(t)\|_E^2 := \frac{\mu}{2}\big(\lambda_+^2(t)+\lambda_-^2(t)\big) + \frac{1}{2}\langle \mathcal{L}_\omega\Gamma(t),\Gamma(t)\rangle_{H^{-1},H^1}. \tag{4.70}$$

Then, we see from (4.69) that

$$\mathcal{H}(\psi)-\mathcal{H}(\Phi_\omega)+\frac{\mu}{2}\big(\lambda_+(t)+\lambda_-(t)\big)^2-\|\eta(t)\|_E^2=O(\,\|\eta(t)\|_{H^1}^{\min\{3,p+1\}}\,). \tag{4.71}$$

LEMMA 4.0.3. *The function* Γ *in the decomposition* (4.66) *satisfies*

$$\|\Gamma(t)\|_{H^1}\lesssim\|\eta(t)\|_{H^1} \tag{4.72}$$

for all $t\in I_{\max}$. *Moreover, we have*

$$\|\eta(t)\|_{H^1}\sim\|\eta(t)\|_E. \tag{4.73}$$

Here, the implicit constants in (4.72) *and* (4.73) *depend on* ω *as well as* d *and* p.

PROOF OF LEMMA 4.0.3. We see from (4.52) that

$$\begin{aligned}\|\Gamma(t)\|_{H^1}^2&=\|\eta(t)-\lambda_+(t)\mathfrak{U}_+-\lambda_-(t)\mathfrak{U}_-\|_{H^1}^2\\&\lesssim\|\eta(t)\|_{H^1}^2+\lambda_+^2(t)\|\mathfrak{U}_+\|_{H^1}^2+\lambda_-^2(t)\|\mathfrak{U}_-\|_{H^1}^2\\&\lesssim(1+2\|\mathfrak{U}_+\|_{H^1}^4)\|\eta(t)\|_{H^1}^2,\end{aligned} \tag{4.74}$$

which proves (4.72).

Next, we shall prove (4.73). We see from (4.66), $\mathfrak{U}_-=\overline{\mathfrak{U}_+}$ (see (4.43)), Lemma 4.0.2 and the definition (4.70) that

$$\begin{aligned}\|\eta(t)\|_{H^1}^2&=\|\lambda_+(t)\mathfrak{U}_++\lambda_-(t)\mathfrak{U}_-+\Gamma(t)\|_{H^1}^2\\&\lesssim\lambda_+^2(t)\|\mathfrak{U}_+\|_{H^1}^2+\lambda_-^2(t)\|\mathfrak{U}_-\|_{H^1}^2+\|\Gamma(t)\|_{H^1}^2\\&\lesssim\|\mathfrak{U}_+\|_{H^1}^2(\lambda_+^2(t)+\lambda_-^2(t))+\langle\mathcal{L}_\omega\Gamma(t),\Gamma(t)\rangle_{H^{-1},H^1}\\&\lesssim\max\{1,\mu^{-1}\}(1+\|\mathfrak{U}_+\|_{H^1}^2)\|\eta(t)\|_E^2.\end{aligned} \tag{4.75}$$

Moreover, it follows from (4.52), Lemma 4.0.2 and (4.74) that

$$\begin{aligned}\|\eta(t)\|_E^2&\lesssim\mu\|\mathfrak{U}_+\|_{L^2}^2\|\eta(t)\|_{L^2}^2+\|\Gamma(t)\|_{H^1}^2\\&\lesssim\mu\|\mathfrak{U}_+\|_{L^2}^2\|\eta(t)\|_{H^1}^2+\big(1+2\|\mathfrak{U}_+\|_{H^1}^4\big)\|\eta(t)\|_{H^1}^2.\end{aligned} \tag{4.76}$$

Putting (4.75) and (4.76) together, we find that (4.73) holds. □

4.0.5. Distance function from the ground state. Our aim here is to introduce a distance function from the ground state Φ_ω by using the linearized energy norm (4.70) (cf. [**25**]). Throughout this subsection, we assume that $d\ge 3$, $1+\frac{4}{d}<p<2^*-1$ and $0<\omega<\omega_2$, where ω_2 is the frequency given by Proposition 4.0.1.

We see from (4.71), and (4.73) in Lemma 4.0.3 that there exists a constant $\delta_E(\omega)>0$ with the following property: for any solution ψ to (NLS) satisfying $\mathcal{M}(\psi)=\mathcal{M}(\Phi_\omega)$ and any $t\in I_{\max}(\psi)$ for which $\|\eta(t)\|_E\le 4\delta_E(\omega)$,

$$\Big|\,\mathcal{H}(\psi(t))-\mathcal{H}(\Phi_\omega)+\frac{\mu}{2}\big(\lambda_+(t)+\lambda_-(t)\big)^2-\|\eta(t)\|_E^2\,\Big|\le\frac{\|\eta(t)\|_E^2}{10}, \tag{4.77}$$

where $\eta(t)$, $\lambda_+(t)$ and $\lambda_-(t)$ are the functions appearing in the decomposition for ψ of the form (4.1) with (4.66) (see also (4.48), (4.53) and (4.54)). Regarding the initial data of a solution to (NLS) as a general function in $H^1(\mathbb{R}^d)$, we find that the following fact holds:

PROPOSITION 4.0.4. *There exists a constant $\delta_E(\omega) > 0$ with the following property: let u be a function in $H^1(\mathbb{R}^d)$ satisfying $\mathcal{M}(u) = \mathcal{M}(\Phi_\omega)$. Consider the decomposition of the form*

(4.78)

$$u = e^{i\theta[u]}(\Phi_\omega + \eta[u]), \quad \Omega(e^{-i\theta[u]}u, \partial_\omega \Phi_\omega) \equiv 0, \quad (e^{-i\theta[u]}u, \partial_\omega \Phi_\omega)_{L^2_{real}} < 0,$$

(4.79)

$$\eta[u] = \lambda_+[u]\mathfrak{U}_+ + \lambda_-[u]\mathfrak{U}_- + \Gamma[u].$$

Define $\|\eta[u]\|_E$ by

$$\|\eta[u]\|_E^2 := \frac{\mu}{2}(\lambda_+^2[u] + \lambda_-^2[u]) + \frac{1}{2}\langle \mathcal{L}_\omega \Gamma[u], \Gamma[u]\rangle_{H^{-1},H^1}. \tag{4.80}$$

Furthermore, assume that

$$\|\eta[u]\|_E \le 4\delta_E(\omega). \tag{4.81}$$

Then,

$$\left| \mathcal{H}(u) - \mathcal{H}(\Phi_\omega) + \frac{\mu}{2}\big(\lambda_+[u] + \lambda_-[u]\big)^2 - \|\eta[u]\|_E^2 \right| \le \frac{\|\eta[u]\|_E^2}{10}. \tag{4.82}$$

Now, we introduce a distance function d_ω. To this end, fix a non-increasing smooth function χ on $[0, \infty)$ such that

$$\chi(r) = \begin{cases} 1 & \text{if} \quad 0 \le r \le 1, \\ 0 & \text{if} \quad r \ge 2. \end{cases} \tag{4.83}$$

Moreover, we define

$$H^1_\omega(\mathbb{R}^d) := \{u \in H^1(\mathbb{R}^d) \colon \mathcal{M}(u) = \mathcal{M}(\Phi_\omega)\}. \tag{4.84}$$

Then, we define a function $d_\omega \colon H^1_\omega(\mathbb{R}^d) \to [0, \infty)$ by

$$d_\omega(u)^2 := \|\eta[u]\|_E^2 + \chi\Big(\frac{\|\eta[u]\|_E}{2\delta_E(\omega)}\Big) C_\omega(u), \tag{4.85}$$

where $\delta_E(\omega)$ is the constant given by Proposition 4.0.4, and

$$C_\omega(u) := \mathcal{H}(u) - \mathcal{H}(\Phi_\omega) + \frac{\mu}{2}\big(\lambda_+[u] + \lambda_-[u]\big)^2 - \|\eta[u]\|_E^2. \tag{4.86}$$

We rephrase (4.82) as follows: if $\|\eta[u]\|_E \le 4\delta_E(\omega)$, then

$$|C_\omega(u)| \le \frac{\|\eta[u]\|_E^2}{10}. \tag{4.87}$$

Now, we consider a solution ψ to (NLS) satisfying $\mathcal{M}(\psi) = \mathcal{M}(\Phi_\omega)$. In the decomposition (4.1) with (4.66), it is convenient to introduce new parameters $\lambda_1(t)$ and $\lambda_2(t)$ defined by

$$\lambda_1(t) := \frac{\lambda_+(t) + \lambda_-(t)}{2}, \qquad \lambda_2(t) := \frac{\lambda_+(t) - \lambda_-(t)}{2}. \tag{4.88}$$

We see from (4.62) and (4.63) that

$$\frac{d\lambda_1}{dt}(t) = \mu\lambda_2(t) + \frac{1}{2}(\Big\{\frac{d\theta}{dt}(t) - \omega\Big\}\eta(t) - N_\omega(\eta(t)), \mathfrak{U}_+ - \mathfrak{U}_-)_{L^2_{real}}, \tag{4.89}$$

$$\frac{d\lambda_2}{dt}(t) = \mu\lambda_1(t) - \frac{1}{2}(\Big\{\frac{d\theta}{dt}(t) - \omega\Big\}\eta(t) - N_\omega(\eta(t)), \mathfrak{U}_+ + \mathfrak{U}_-)_{L^2_{real}}. \tag{4.90}$$

An important property of the distance function $d_\omega(\psi(t))$ is the following:

LEMMA 4.0.5. *Assume that there exists an interval I on which*

$$\sup_{t\in I} d_\omega(\psi(t)) \le \delta_E(\omega). \tag{4.91}$$

Then, all of the following hold for all $t \in I$:

$$\frac{1}{2}\|\eta(t)\|_E^2 \le d_\omega(\psi(t))^2 \le \frac{3}{2}\|\eta(t)\|_E^2, \tag{4.92}$$

$$d_\omega(\psi(t))^2 = \mathcal{H}(\psi) - \mathcal{H}(\Phi_\omega) + 2\mu\lambda_1^2(t), \tag{4.93}$$

$$\frac{d}{dt}d_\omega(\psi(t))^2 = 4\mu^2\lambda_1(t)\lambda_2(t) + 4\mu\lambda_1(t)\Omega\big(\Big\{\frac{d\theta}{dt}(t) - \omega\Big\}\eta(t) - N_\omega(\eta(t)), f_2\big). \tag{4.94}$$

Furthermore, if

$$\mathcal{S}_\omega(\psi) < m_\omega + \frac{1}{2}d_\omega(\psi(t))^2 \tag{4.95}$$

holds for all $t \in I$, then

$$d_\omega(\psi(t)) \sim |\lambda_1(t)| \tag{4.96}$$

for all $t \in I$, where the implicit constant depends on ω as well as d and p.

PROOF OF LEMMA 4.0.5. First, we shall show that for any $t \in I$,

$$\|\eta(t)\|_E \le 4\delta_E(\omega). \tag{4.97}$$

Suppose for contradiction that $\|\eta(t_0)\|_E > 4\delta_E(\omega)$ for some $t_0 \in I$. Then, it follows from the definition (4.85) that $d_\omega(\psi(t_0)) = \|\eta(t_0)\|_E \ge 4\delta_E(\omega)$. However, this contradicts the assumption (4.91). Thus, we have proved (4.97). Using (4.87) and (4.97), we can verify (4.92). Indeed,

$$\begin{aligned}\frac{1}{2}\|\eta(t)\|_E^2 &\le \|\eta(t)\|_E^2 - \big|C_\omega(\psi(t))\big| \\ &\le d_\omega(\psi(t))^2 \le \|\eta(t)\|_E^2 + \big|C_\omega(\psi(t))\big| \le \frac{3}{2}\|\eta(t)\|_E^2.\end{aligned} \tag{4.98}$$

Next, we shall derive the equation (4.93). We see from (4.91) and (4.92) that

$$\|\eta(t)\|_E^2 \le 2d_\omega(\psi(t))^2 \le 2\delta_E(\omega)^2. \tag{4.99}$$

Furthermore, it follows from the definition of $d_\omega(\psi(t))$ (see (4.85) and (4.86)) that

$$d_\omega(\psi(t))^2 = \|\eta(t)\|_E^2 + C_\omega(\psi(t)) = \mathcal{H}(\psi) - \mathcal{H}(\Phi_\omega) + \frac{\mu}{2}\big(\lambda_+(t) + \lambda_-(t)\big)^2, \tag{4.100}$$

so that (4.93) holds.

The equation (4.94) follows from (4.93) and (4.89).

Finally, we shall prove (4.96). We see from the definition of $C_\omega(\psi(t))$ (see (4.86)), (4.97) and (4.87) that

$$\begin{aligned}\|\eta(t)\|_E^2 &= \mathcal{H}(\psi) - \mathcal{H}(\Phi_\omega) + \frac{\mu}{2}\big(\lambda_+(t) + \lambda_-(t)\big)^2 - C_\omega(\psi(t)) \\ &\le \mathcal{H}(\psi) - \mathcal{H}(\Phi_\omega) + \frac{\mu}{2}\big(\lambda_+(t) + \lambda_-(t)\big)^2 + \frac{\|\eta(t)\|_E^2}{10},\end{aligned} \tag{4.101}$$

so that

$$\frac{9}{10}\|\eta(t)\|_E^2 \le \mathcal{H}(\psi) - \mathcal{H}(\Phi_\omega) + 2\mu\lambda_1(t)^2. \tag{4.102}$$

Moreover, it follows from the assumptions (4.55) and (4.95), and $m_\omega = \mathcal{S}_\omega(\Phi_\omega)$ that

$$\mathcal{H}(\psi) - \mathcal{H}(\Phi_\omega) = \mathcal{S}_\omega(\psi) - \mathcal{S}_\omega(\Phi_\omega) < \frac{1}{2}d_\omega(\psi(t))^2. \tag{4.103}$$

Putting (4.92), (4.102) and (4.103) together, we obtain that

$$d_\omega(\psi(t))^2 \le \frac{3}{2}\|\eta(t)\|_E^2 < \frac{5}{6}d_\omega(\psi(t))^2 + \frac{10}{3}\mu\lambda_1(t)^2. \tag{4.104}$$

Hence, we have

$$d_\omega(\psi(t))^2 < 200\mu\lambda_1(t)^2. \tag{4.105}$$

On the other hand, we see from (4.52), Lemma 4.0.3 and (4.92) that

$$\mu\lambda_1(t)^2 \le 2\mu\big(\lambda_+^2(t) + \lambda_-^2(t)\big) \le 4\mu\|\mathfrak{U}_+\|_{L^2}^2\|\eta(t)\|_{L^2}^2 \lesssim d_\omega(\psi(t))^2, \tag{4.106}$$

where the implicit constant depends on μ and $\|\mathfrak{U}_+\|_{L^2}^2$. Combining (4.105) and (4.106), we obtain (4.96). □

4.0.6. Fundamental properties of the eigenfunctions. In connection with λ_1 and λ_2 (see (4.88)), we introduce the real-valued functions f_1 and f_2:

$$f_1 := \frac{\mathfrak{U}_+ + \mathfrak{U}_-}{2} = \Re[\mathfrak{U}_+], \qquad f_2 := \frac{\mathfrak{U}_+ - \mathfrak{U}_-}{2i} = \Im[\mathfrak{U}_+]. \tag{4.107}$$

We shall observe the properties of f_1 and f_2. Throughout this subsection, we assume that $d \ge 3$, $1 + \frac{4}{d} < p < 2^* - 1$, $0 < \omega < \omega_2$ and (4.55), where ω_2 is the constant given by Proposition 4.0.1.

First, we note that the decomposition (4.66) of $\eta(t)$ is expressed as follows in terms of the functions f_1 and f_2 defined by (4.107):

$$\eta(t) = 2\lambda_1(t)f_1 + 2i\lambda_2(t)f_2 + \Gamma(t). \tag{4.108}$$

We see from (4.6) and $\mathcal{L}_\omega\mathfrak{U}_+ = i\mu\mathfrak{U}_+$ that

$$L_{\omega,+}f_1 = -\mu f_2, \qquad L_{\omega,-}f_2 = \mu f_1. \tag{4.109}$$

Furthermore, it follows from the elliptic regularity for (4.109) that

$$\|f_1\|_{L^\infty} + \|f_2\|_{L^\infty} < \infty. \tag{4.110}$$

LEMMA 4.0.6. *We have that* $f_2 \notin \operatorname{span}\{\Phi_\omega\} = \operatorname{Ker} L_{\omega,-}$.

PROOF OF LEMMA 4.0.6. Suppose for a contradiction that $f_2 = k\Phi_\omega$ for some $k \neq 0$. Then, it follows from (4.8) and (4.107) that

$$0 = k\mathcal{L}_\omega(i\Phi_\omega) = \mathcal{L}_\omega(if_2) = \frac{1}{2}\mathcal{L}_\omega(\mathfrak{U}_+ - \mathfrak{U}_-) = \frac{1}{2}i\mu\mathfrak{U}_+ + \frac{1}{2}i\mu\mathfrak{U}_- = i\mu f_1. \tag{4.111}$$

Furthermore, we see from (4.109) and (4.111) that

$$-\mu f_2 = L_{\omega,+}f_1 = 0. \tag{4.112}$$

Thus, $f_1 = f_2 \equiv 0$. This is a contradiction. □

LEMMA 4.0.7. *We have the following orthogonalities:*

$$(\Phi_\omega, f_1)_{L^2} = (\partial_\omega \Phi_\omega, f_2)_{L^2} = 0. \tag{4.113}$$

Furthermore, we have

$$(f_1, f_2)_{L^2} > 0 \tag{4.114}$$

and

$$(\Phi_\omega, f_2)_{L^2} \neq 0. \tag{4.115}$$

PROOF OF LEMMA 4.0.7. Since $L_{\omega,-}\Phi_\omega = 0$ and $L_{\omega,-}$ is self-adjoint in $L^2(\mathbb{R}^d)$, we find from (4.109) that

$$(\Phi_\omega, f_1)_{L^2} = \mu^{-1}\langle L_{\omega,-}f_2, \Phi_\omega\rangle_{H^{-1},H^1} = \mu^{-1}(L_{\omega,-}\Phi_\omega, f_2)_{L^2} = 0. \tag{4.116}$$

On the other hand, it follows from (4.109), (4.9) and the self-adjointness of $L_{\omega,+}$ in $L^2(\mathbb{R}^d)$ that

$$\begin{aligned}(\partial_\omega \Phi_\omega, f_2)_{L^2} &= -\mu^{-1}\langle L_{\omega,+}f_1, \partial_\omega\Phi_\omega\rangle_{H^{-1},H^1} \\ &= -\mu^{-1}(L_{\omega,+}\partial_\omega\Phi_\omega, f_1)_{L^2} = \mu^{-1}(\Phi_\omega, f_1)_{L^2},\end{aligned} \tag{4.117}$$

which together with (4.116) proves (4.113).

Next, we shall prove (4.114). It follows from (4.109) that

$$(f_1, f_2)_{L^2} = \mu^{-1}\langle L_{\omega,-}f_2, f_2\rangle_{H^{-1},H^1}. \tag{4.118}$$

Since $f_2 \notin \operatorname{Ker} L_{\omega,-}$ (see Lemma 4.0.6), Lemma B.0.1 together with (4.118) shows the desired result.

Finally, we prove (4.115). Suppose for contradiction that $(\Phi_\omega, f_2)_{L^2} = 0$. Then, we see from (4.7), the self-adjointness of $L_{\omega,+}$ and (4.109) that

$$((p-1)\Phi_\omega^p + (2^*-2)\Phi_\omega^{2^*-1}, f_1)_{L^2} = -(L_{\omega,+}\Phi_\omega, f_1)_{L^2} = \mu(\Phi_\omega, f_2)_{L^2} = 0. \tag{4.119}$$

Hence, it follows from Lemma B.0.3 that $\langle L_{\omega,+}f_1, f_1\rangle_{H^{-1},H^1} \geq 0$. However, it must follow from (4.109) and (4.114) that

$$\langle L_{\omega,+}f_1, f_1\rangle_{H^{-1},H^1} = -\mu(f_2, f_1)_{L^2} < 0. \tag{4.120}$$

This is a contradiction. Thus, we find that (4.115) holds. ☐

The relation (4.115) in Lemma 4.0.7 allows us to choose f_2 so that

$$(\Phi_\omega, f_2)_{L^2} < 0. \tag{4.121}$$

In our analysis, the frequency ω varies. In particular, we need to take $\omega \to 0$. Hence, we have to pay attention to the dependence of ω. Such a bother does not appear in the scale invariant cases such as (1.18) (see [**25**]). The following lemma plays an important role to prove the ejection lemma (Lemma 5.0.2):

LEMMA 4.0.8. *For any constant $C > 0$, there exists $\omega(C) > 0$ such that for any $\omega \in (0, \omega(C))$,*

$$\mu|(\Phi_\omega, f_2)_{L^2}| \geq C|(\Phi_\omega^{2^*-1}, f_1)_{L^2}|. \tag{4.122}$$

PROOF OF LEMMA 4.0.8. We use the notation μ_ω, $f_{1,\omega}$ and $f_{2,\omega}$ instead of μ, f_1 and f_2 in order to emphasise the dependence on ω.

Suppose for contradiction that there exists a constant $C_0 > 0$ with the following property: for any number n, there exists $\omega_n \in (0, \frac{1}{n})$ such that

$$\mu_{\omega_n}|(\Phi_{\omega_n}, f_{2,\omega_n})_{L^2}| < C_0|(\Phi_{\omega_n}^{2^*-1}, f_{1,\omega_n})_{L^2}|, \tag{4.123}$$

where f_{1,ω_n} and f_{2,ω_n} are functions given by (4.107) for ω_n. We consider the functions $f_{1,n}$ and $f_{2,n}$ defined by

$$f_{1,n} := \omega_n^{-s_p} T_{\omega_n} f_{1,\omega_n}, \qquad f_{2,n} := \omega_n^{-s_p} T_{\omega_n} f_{2,\omega_n}. \tag{4.124}$$

It follows from (4.109) that

$$\widetilde{L}_{\omega_n,+} f_{1,n} = -\mu_{\omega_n} \omega_n^{-1} f_{2,n}, \quad \widetilde{L}_{\omega_n,-} f_{2,n} = \mu_{\omega_n} \omega_n^{-1} f_{1,n}, \tag{4.125}$$

where $\widetilde{L}_{\omega,+}$ and $\widetilde{L}_{\omega,-}$ are the operators defined by (1.59) and (1.60), respectively. We put

$$g_n := \frac{f_{1,n}}{\|f_{1,n}\|_{H^1}}, \qquad h_n := \frac{f_{2,n}}{\|f_{1,n}\|_{H^1}}. \tag{4.126}$$

Then, we can take a real-valued function g in $H^1(\mathbb{R}^d)$ such that

$$\lim_{n\to\infty} g_n = g \qquad \text{weakly in } H^1(\mathbb{R}^d). \tag{4.127}$$

We shall show that g is non-trivial. Suppose for contradiction that g was trivial. Then, it follows from Lemma 4.0.7, (4.125), Proposition 2.0.3 and (4.127) that

$$\begin{aligned}
0 &\ge \lim_{n\to\infty} \frac{-\omega_n^{-s_p-1}\mu_{\omega_n}}{\|f_{1,n}\|_{H^1}^2} (f_{1,\omega_n}, f_{2,\omega_n})_{L^2} = \lim_{n\to\infty} \frac{-\omega_n^{-1}\mu_{\omega_n}}{\|f_{1,n}\|_{H^1}^2} (f_{1,n}, f_{2,n})_{L^2} \\
&= \lim_{n\to\infty} \frac{1}{\|f_{1,n}\|_{H^1}^2} \langle \widetilde{L}_{\omega_n,+} f_{1,n}, f_{1,n}\rangle_{H^{-1},H^1} \\
&= 1 - \lim_{n\to\infty} \int_{\mathbb{R}^d} \Big\{ p(T_{\omega_n}\Phi_{\omega_n})^{p-1}|g_n|^2 + \omega_n^{\frac{2^*-(p+1)}{p-1}} (2^*-1)(T_{\omega_n}\Phi_{\omega_n})^{\frac{4}{d-2}}|g_n|^2 \Big\}\, dx \\
&= 1.
\end{aligned} \tag{4.128}$$

This is a contradiction, and therefore g is non-trivial.

Next, we shall show that there exists a constant $C_0 > 0$ such that for any number n and any real-valued function $f \in H^1(\mathbb{R}^d)$ with $(f, T_{\omega_n}\Phi_{\omega_n})_{L^2} = 0$,

$$\langle \widetilde{L}_{\omega_n,-} f, f\rangle_{H^{-1},H^1} \ge C_0 \|f\|_{H^1}^2. \tag{4.129}$$

Suppose for contradiction that we could take a subsequence of $\{\omega_n\}$ (still denoted by the same symbol) and a sequence $\{f_n\}$ in $H^1(\mathbb{R}^d)$ such that

$$\lim_{n\to\infty} \langle \widetilde{L}_{\omega_n,-} f_n, f_n\rangle_{H^{-1},H^1} = 0, \qquad \|f_n\|_{H^1} = 1, \qquad (f_n, T_{\omega_n}\Phi_{\omega_n})_{L^2} = 0. \tag{4.130}$$

Furthermore, we can take $f_0 \in H^1(\mathbb{R}^d)$ such that $\lim_{n\to\infty} f_n = f_0$ weakly in $H^1(\mathbb{R}^d)$. Then, we see from the weak lower semicontinuity and Proposition 2.0.3 that

$$\begin{aligned}
0 &= \lim_{n\to\infty} \langle \widetilde{L}_{\omega_n,-} f_n, f_n\rangle_{H^{-1},H^1} \\
&= \lim_{n\to\infty} \Big\{ \|f_n\|_{H^1}^2 - \int (T_{\omega_n}\Phi_{\omega_n})^{p-1}|f_n|^2 dx - \omega_n^{\frac{2^*-(p+1)}{p-1}} \int (T_{\omega_n}\Phi_{\omega_n})^{\frac{4}{d-2}}|f_n|^2 dx \Big\} \\
&\ge \langle L_-^{\dagger} f_0, f_0\rangle_{H^{-1},H^1}.
\end{aligned} \tag{4.131}$$

Moreover, it follows from Proposition 2.0.3 and the hypothesis (4.130) that

$$(f_0, U)_{L^2} = \lim_{n\to\infty} (f_n, T_{\omega_n}\Phi_{\omega_n})_{L^2} = 0. \tag{4.132}$$

Hence, we conclude from (4.131) and the positivity of $L_-^\dagger$ together with (4.132) that

$$0 \ge \langle L_-^\dagger f_0, f_0\rangle \gtrsim \|f_0\|_{H^1}^2, \tag{4.133}$$

so that f_0 is trivial. However, the same argument as (4.128) yields that f_0 is non-trivial. Thus, we arrive at a contradiction and therefore (4.129) holds.

We shall show that $\{h_n\}$ is bounded in $H^1(\mathbb{R}^d)$. We see from (4.129) and (4.125) that

$$\begin{aligned} C_0\|f_{2,n}\|_{H^1}^2 &\le \langle \widetilde{L}_{\omega_n,-}f_{2,n}, f_{2,n}\rangle_{H^{-1},H^1} = \mu_{\omega_n}\omega_n^{-1}(f_{1,n}, f_{2,n})_{L^2} \\ &= -\langle \widetilde{L}_{\omega_n,+}f_{1,n}, f_{1,n}\rangle_{H^{-1},H^1} \le C\|f_{1,n}\|_{H^1}^2. \end{aligned} \tag{4.134}$$

Dividing the both sides above by $\|f_{1,n}\|_{H^1}^2$, we find that $\{h_n\}$ is bounded in $H^1(\mathbb{R}^d)$.

We shall show that the sequence $\{\mu_{\omega_n}\omega_n^{-1}\}$ is bounded. It follows from (4.125) that

$$\left|\mu_{\omega_n}\omega_n^{-1}\right| = \frac{|\langle \widetilde{L}_{\omega_n,-}f_{2,n}, f_{1,n}\rangle_{H^{-1},H^1}|}{\|f_{1,n}\|_{L^2}^2}. \tag{4.135}$$

Furthermore, it follows from (4.127) and the boundedness of $\{g_n\}$ and $\{h_n\}$ in $H^1(\mathbb{R}^d)$ that

(4.136)

$$\left|\mu_{\omega_n}\omega_n^{-1}\right| \le \frac{\sup_{n\in\mathbb{N}} |\langle \widetilde{L}_{\omega_n,-}h_n, g_n\rangle_{H^{-1},H^1}|}{\liminf_{n\to\infty}\|g_n\|_{L^2}^2} \lesssim \frac{\sup_{n\in\mathbb{N}}(\|g_n\|_{H^1}^2 + \|h_n\|_{H^1}^2)}{\|g\|_{L^2}^2} \lesssim 1.$$

Thus, we find that $\{\mu_{\omega_n}\omega_n^{-1}\}$ is bounded.

Since $\{h_n\}$ is bounded in $H^1(\mathbb{R}^d)$, there exists $h \in H^1(\mathbb{R}^d)$ such that

$$\lim_{n\to\infty} h_n = h \qquad \text{weakly in } H^1(\mathbb{R}^d). \tag{4.137}$$

Moreover, we can take a subsequence of $\{\mu_{\omega_n}\omega_n^{-1}\}$ (still denoted by the same symbol) and $\nu_* \in [0,\infty)$ such that $\lim_{n\to\infty}\mu_{\omega_n}\omega_n^{-1} = \nu_*$. We shall show that $\nu_* \neq 0$ and h is non-trivial. Recall here that μ_{ω_n} is a positive eigenvalue of $-i\mathcal{L}_{\omega_n}$. We find from the proof of Proposition 4.0.1 (see (4.15) and (4.17)) that

$$-\mu_{\omega_n}^2 = \inf\left\{ \frac{\langle L_{\omega_n,+}u, u\rangle_{H^{-1},H^1}}{((L_{\omega_n,-})^{-1}u, u)_{L^2_{real}}} : u \in H^1(\mathbb{R}^d), (u, \Phi_{\omega_n})_{L^2_{real}} = 0 \right\}. \tag{4.138}$$

Furthermore, it follows from the estimates (4.21) and (4.22) in the proof of Proposition 4.0.1 shows that

$$-\nu_*^2 = \lim_{n\to\infty} \omega_n^{-2}(-\mu_{\omega_n}^2) = \frac{\langle L_+^\dagger \Pi\Lambda U, \Pi\Lambda U\rangle_{H^{-1},H^1}}{((L_-^\dagger)^{-1}\Pi\Lambda U, \Pi\Lambda U)_{L^2_{real}}} < 0, \tag{4.139}$$

where Π is the projection given by (4.20). Thus, we find that $\nu_* \neq 0$. Next, suppose for contradiction that h was trivial. Then, we see from (4.125) and (4.127) that for any $\varphi \in C_0^\infty(\mathbb{R}^d)$,

$$0 = \lim_{n\to\infty} \langle \widetilde{L}_{\omega_n,-}h_n, \varphi\rangle_{H^{-1},H^1} = \lim_{n\to\infty} (\mu_{\omega_n}\omega_n^{-1}g_n, \varphi)_{L^2_{real}} = (\nu_* g, \varphi)_{L^2_{real}}. \tag{4.140}$$

However, this contradicts that $\nu_* \neq 0$ and g is non-trivial. Thus, we find that h is non-trivial.

Note that g and h satisfies that

$$L_+^\dagger g = -\nu_* h, \qquad L_-^\dagger h = \nu_* g. \tag{4.141}$$

Then, applying the same proof of (4.115) (use Lemma 2.2 in **[9]** instead of Lemma B.0.3), we find that

$$\langle L_+^\dagger U, h\rangle_{H^{-1},H^1} = (p-1)(U^p, h)_{L^2} \neq 0. \tag{4.142}$$

We see from (4.109) and the scaling that

$$-\omega_n^{-1}\mu_{\omega_n}(\Phi_{\omega_n}, f_{2,\omega_n})_{L^2} = \langle \widetilde{L}_{\omega_n,+}T_{\omega_n}\Phi_{\omega_n}, f_{1,n}\rangle_{H^{-1},H^1}. \tag{4.143}$$

Moreover, it follows from Proposition 2.0.3 that

$$\lim_{n\to\infty} \widetilde{L}_{\omega_n,+}T_{\omega_n}\Phi_{\omega_n} = L_+^\dagger U \qquad \text{strongly in } H^{-1}(\mathbb{R}^d). \tag{4.144}$$

This together with (4.137) yields

$$\lim_{n\to\infty} \langle \widetilde{L}_{\omega_n,+}T_{\omega_n}\Phi_{\omega_n}, h_n\rangle_{H^{-1},H^1} = \langle L_+^\dagger U, h\rangle_{H^{-1},H^1}. \tag{4.145}$$

We find from (4.143), (4.144) and (4.142) that

$$\begin{aligned}\lim_{n\to\infty} \|f_{1,n}\|_{H^1}^{-1}\omega_n^{-1}\mu_{\omega_n}\big|(\Phi_{\omega_n}, f_{2,\omega_n})_{L^2}\big| &= \lim_{n\to\infty} \big|\langle \widetilde{L}_{\omega_n,+}T_{\omega_n}\Phi_{\omega_n}, h_n\rangle_{H^{-1},H^1}\big| \\ &= \big|\langle L_+^\dagger U, h\rangle_{H^{-1},H^1}\big| \gtrsim 1,\end{aligned} \tag{4.146}$$

On the other hand, it follows from Proposition 2.0.3 and (4.127) that

$$\begin{aligned}\lim_{n\to 0} \|f_{1,n}\|_{H^1}^{-1}\omega_n^{-\frac{2^*-2}{p-1}}\big|(\Phi_{\omega_n}^{2^*-1}, f_{1,,\omega_n})_{L^2}\big| &= \lim_{n\to 0}\big|\big((T_{\omega_n}\Phi_{\omega_n})^{2^*-1}, g_n\big)_{L^2}\big| \\ &= \big|(U^{2^*-1}, g)_{L^2}\big| \lesssim 1.\end{aligned} \tag{4.147}$$

Since $\frac{2^*-2}{p-1} > 1$, we conclude from (4.146) and (4.147) that for any sufficiently large number n,

$$\mu_{\omega_n}\big|(\Phi_{\omega_n}, f_{2,,\omega_n})_{L^2}\big| \gtrsim \omega_n^{-\frac{2^*-2}{p-1}+1}\big|(\Phi_{\omega_n}^{2^*-1}, f_{1,,\omega_n})_{L^2}\big| \gg C_0\big|(\Phi_{\omega_n}^{2^*-1}, f_{1,\omega_n})_{L^2}\big|, \tag{4.148}$$

where C_0 is the constant given in the hypothesis (4.123). However, this contradicts (4.123). Thus, the desired result (4.122) holds. □

Now, we see from (4.47) and (4.57) that

$$(\Phi_\omega, \Gamma(t))_{L^2_{real}} = -\mathcal{M}(\eta(t)). \tag{4.149}$$

Moreover, we see from (4.7) and (4.8) that

$$\begin{aligned}&2\omega\Phi_\omega - 2\Delta\Phi_\omega - \frac{d(p-1)}{2}\Phi_\omega^p - 2^*\Phi_\omega^{2^*-1} \\ &= (2-s_p)L_{\omega,-}\Phi_\omega + s_pL_{\omega,+}\Phi_\omega - (1-s_p)(2^*-2)\Phi_\omega^{2^*-1} \\ &= s_pL_{\omega,+}\Phi_\omega - (1-s_p)(2^*-2)\Phi_\omega^{2^*-1} = -s_p(p-1)\Phi_\omega^p - (2^*-2)\Phi_\omega^{2^*-1}.\end{aligned} \tag{4.150}$$

Furthermore, this together with (4.108), (4.109) and (4.149) shows that

(4.151)

$$
\begin{aligned}
&\mathcal{K}'(\Phi_\omega)\eta(t) \\
&= 2\langle \omega\Phi_\omega - \Delta\Phi_\omega - \frac{d(p-1)}{4}\Phi_\omega^p - \frac{2^*}{2}\Phi_\omega^{2^*-1}, \eta(t)\rangle_{H^{-1},H^1} - 2\omega(\Phi_\omega, \eta(t))_{L^2_{real}} \\
&= 2s_p\lambda_1(t)(L_{\omega,+}\Phi_\omega, f_1)_{L^2} - 2(1-s_p)(2^*-2)\lambda_1(t)(\Phi_\omega^{2^*-1}, f_1)_{L^2} \\
&\quad - s_p((p-1)\Phi_\omega^p + (2^*-2)\Phi_\omega^{2^*-1}, \Gamma(t))_{L^2_{real}} \\
&\quad - (1-s_p)(2^*-2)(\Phi_\omega^{2^*-1}, \Gamma(t))_{L^2_{real}} + \omega\mathcal{M}(\eta(t)) \\
&= -2\mu s_p\lambda_1(t)(\Phi_\omega, f_2)_{L^2} - 2(1-s_p)(2^*-2)\lambda_1(t)(\Phi_\omega^{2^*-1}, f_1)_{L^2} \\
&\quad - s_p(p-1)(\Phi_\omega^p, \Gamma(t))_{L^2_{real}} - (2^*-2)(\Phi_\omega^{2^*-1}, \Gamma(t))_{L^2_{real}} + \omega\mathcal{M}(\eta(t)).
\end{aligned}
$$

CHAPTER 5

Ejection lemma

Let ω_2 denote the frequency given by Proposition 4.0.1 throughout this section. Furthermore, for a given $\omega \in (0, \omega_2)$, $\delta_E(\omega)$ denotes the constant given by Proposition 4.0.4.

We see from the argument in the previous section that any solution ψ satisfying $\mathcal{M}(\psi) = \mathcal{M}(\Phi_\omega)$ has the decomposition of the form (4.1) with (4.48), (4.53) and (4.54). Recall that $d_\omega \colon H^1_\omega(\mathbb{R}^d) \to [0, \infty)$ denotes the distance function defined by (4.85).

The following lemma is a criterion of continuation for the solutions to (NLS) in terms of the ground state:

LEMMA 5.0.1. *Assume $d \geq 3$ and $1+\frac{4}{d} < p < 2^*-1$. Then, for any $\omega \in (0, \omega_2)$, there exists $\delta_0(\omega) \in (0, \delta_E(\omega))$ such that if ψ is a solution to* (NLS) *satisfying*

$$\mathcal{M}(\psi) = \mathcal{M}(\Phi_\omega), \tag{5.1}$$

then ψ extends as long as $d_\omega(\psi(t)) \leq \delta_0(\omega)$.

PROOF OF LEMMA 5.0.1. We prove the claim by contradiction. Hence, suppose to the contrary that there exists $\omega \in (0, \omega_2)$ with the following property: for any $\delta_0 \in (0, \delta_E(\omega))$, there exists a solution ψ to (NLS) such that: $\mathcal{M}(\psi) = \mathcal{M}(\Phi_\omega)$; $T_{\max} := \sup I_{\max}(\psi) < \infty$; and $\sup_{t\in[t_0,T_{\max})} d_\omega(\psi(t)) \leq \delta_0$ for some $t_0 \in I_{\max}$. Then, it follows from Lemma 4.0.3 and Lemma 4.0.5 that

$$\sup_{t\in[t_0,T_{\max})} \|\eta(t)\|_{H^1} \lesssim \sup_{t\in[t_0,T_{\max})} d_\omega\big(\psi(t)\big) < \delta_0, \tag{5.2}$$

where the implicit constant depends only on d, p and ω. In particular, we have

$$\sup_{t\in[t_0,T_{\max})} \|\psi(t)\|_{H^1} \lesssim \|\Phi_\omega\|_{H^1} + \delta_0. \tag{5.3}$$

Let $t_1 \in (t_0, T_{\max})$. Then, Strichartz' estimate together with (5.2) shows that

$$\begin{aligned} \|\langle\nabla\rangle e^{i(t-t_1)\Delta}\psi(t_1)\|_{St([t_1,T_{\max}])} &\lesssim \|\langle\nabla\rangle e^{i(t-t_1)\Delta}\Phi_\omega\|_{St([t_1,T_{\max}])} + \|\eta(t_1)\|_{H^1} \\ &\lesssim \|\langle\nabla\rangle e^{i(t-t_1)\Delta}\Phi_\omega\|_{St([t_1,T_{\max}])} + \delta_0. \end{aligned} \tag{5.4}$$

Furthermore, the small-data theory (Lemma D.0.1) together with (5.3) shows that there exists $\delta(\omega) > 0$ such that if

$$\|\langle\nabla\rangle e^{i(t-t_1)\Delta}\psi(t_1)\|_{St([t_1,T_{\max}])} \leq \delta(\omega), \tag{5.5}$$

then the solution ψ exists on $[t_1, T_{\max}]$. Since

$$\lim_{t_1\uparrow T_{\max}} \|\langle\nabla\rangle e^{i(t-t_1)\Delta}\Phi_\omega\|_{St([t_1,T_{\max}])} = 0, \tag{5.6}$$

we see from (5.4) that if $\delta_0 \ll \delta(\omega)$, then ψ extends beyond $T_{\max}$. However, this is a contradiction. Thus, we have proved the lemma. $\square$

Note here that if follows from Lemma 4.0.8 that there exists a frequency $\omega(2^*) \in (0,\omega_2)$ such that for any $\omega \in (0,\omega(2^*))$,

$$\mu s_p \big|(\Phi_\omega, f_2)_{L^2}\big| \geq 2(1-s_p)(2^*-2)\big|(\Phi_\omega^{2^*-1}, f_1)_{L^2}\big|. \tag{5.7}$$

We use this fact in the proof of the ejection lemma below.

LEMMA 5.0.2 (Ejection lemma). *Assume $d \geq 3$ and $1+\frac{4}{d} < p < 2^*-1$. Then, for any $\omega \in (0,\omega(2^*))$, there exist constants $\delta_X \in (0,\delta_0(\omega))$ ($\delta_0(\omega)$ denotes the constant given by Lemma 5.0.1), $A_* > 0$, $B_* > 0$, $C_* > 0$ and $T_* > 0$ with the following properties: for any $t_0 \in \mathbb{R}$ and any solution ψ to* (NLS) *defined around t_0 satisfying*

$$\mathcal{M}(\psi) = \mathcal{M}(\Phi_\omega), \tag{5.8}$$

$$0 < R_0 := d_\omega(\psi(t_0)) < \delta_X, \tag{5.9}$$

$$\mathcal{S}_\omega(\psi) < m_\omega + \frac{R_0^2}{2}, \tag{5.10}$$

we can extend ψ as long as $d_\omega(\psi(t)) \leq \delta_X$. Furthermore, assume that there exists $T > t_0$ such that

$$R_0 \leq \min_{t\in[t_0,T]} d_\omega(\psi(t)), \tag{5.11}$$

and define

$$T_X := \inf\big\{t \in [t_0,T] \colon d_\omega(\psi(t)) = \delta_X\big\}, \tag{5.12}$$

where we interpret $T_X = T$ if $d_\omega(\psi) < \delta_X$ on $[t_0,T]$. Then, for any $t \in [t_0,T_X]$,

$$A_* e^{\mu(t-t_0)} R_0 \leq d_\omega(\psi(t)) \leq B_* e^{\mu(t-t_0)} R_0, \tag{5.13}$$

$$\|\eta(t)\|_{H^1} \sim \mathfrak{s}\lambda_1(t) \sim \mathfrak{s}\lambda_+(t) \sim e^{\mu(t-t_0)} R_0, \tag{5.14}$$

$$|\lambda_-(t)| + \|\Gamma(t)\|_{H^1} \lesssim R_0 + \big(e^{\mu(t-t_0)} R_0\big)^{\frac{\min\{3,p+1\}}{2}}, \tag{5.15}$$

$$\mathfrak{s}\mathcal{K}(\psi(t)) \gtrsim \big(e^{\mu(t-t_0)} - C_*\big) R_0, \tag{5.16}$$

where $\mathfrak{s}$ is either $\mathfrak{s} = 1$ or $\mathfrak{s} = -1$. Moreover, $d_\omega(\psi(t))$ is increasing on the region $\{t \in [t_0,T_X] \colon t_0 + T_* R_0^{\min\{1,p-1\}} \leq t\}$; and

$$\big|d_\omega(\psi(t)) - R_0\big| \lesssim R_0^{\min\{2,p\}} \tag{5.17}$$

on the region $\{t \in [t_0,T_X] \colon t_0 \leq t \leq t_0 + T_* R_0^{\min\{1,p-1\}}\}$.

PROOF OF LEMMA 5.0.2. Let $\delta_X \in (0,\delta_0(\omega))$ be a small constant to be chosen later, and let ψ be a solution satisfying (5.8), (5.9) and (5.10). Since the equation (NLS) is invariant under the time translations, it suffices to consider the case where $t_0 = 0$.

First, we find from Lemma 5.0.1 that ψ extends as long as $d_\omega(\psi(t)) \leq \delta_X$.

Next, we assume (5.11) as well as (5.8), (5.9) and (5.10). Then, it follows from the definition of T_X and $t_0 = 0$ that for any $t \in [0,T_X]$,

$$d_\omega(\psi(t)) \leq \delta_X < \delta_E(\omega). \tag{5.18}$$

Hence, we see from Lemma 4.0.3 and Lemma 4.0.5 that for any $t \in [0, T_X]$,

(5.19)
$$\|\Gamma(t)\|_{H^1} \lesssim \|\eta(t)\|_{H^1} \sim \|\eta(t)\|_E \sim d_\omega(\psi(t)) \sim |\lambda_1(t)|,$$

(5.20)
$$\frac{d}{dt} d_\omega\big(\psi(t)\big)^2 = 4\mu^2\lambda_1(t)\lambda_2(t) + 4\mu\lambda_1(t)\Omega\Big(\Big\{\frac{d\theta}{dt}(t) - \omega\Big\}\eta(t) - N_\omega(\eta(t)), f_2\Big).$$

In particular, we deduce from (4.52), (5.18) and (5.19) that for any $t \in [0, T_X]$,

$$|\lambda_-(t)| + \|\Gamma(t)\|_{H^1} \lesssim \|\eta(t)\|_{H^1} \sim |\lambda_1(t)| \lesssim \delta_X \ll 1. \tag{5.21}$$

Furthermore, we find from (5.11) and (5.19) that for any $t \in [0, T_X]$.

$$0 < R_0 \lesssim |\lambda_1(t)|, \tag{5.22}$$

which together with the continuity of $\lambda_1(t)$ shows that for any $t \in [0, T_X]$

$$\mathfrak{s} := \operatorname{sgn}[\lambda_1(t)] \equiv \operatorname{sgn}[\lambda_1(0)]. \tag{5.23}$$

Now, we further choose δ_X so small that

$$\delta_X \ll \frac{|(\Phi_\omega, \partial_\omega \Phi_\omega)_{L^2}|}{2\|\partial_\omega \Phi_\omega\|_{L^2}}. \tag{5.24}$$

Then, we find from (5.21) that for any $t \in [0, T_X]$,

$$\|\eta(t)\|_{L^2} \le \|\eta(t)\|_{H^1} \ll \min\Big\{1, \frac{|(\Phi_\omega, \partial_\omega \Phi_\omega)_{L^2}|}{2\|\partial_\omega \Phi_\omega\|_{L^2}}\Big\}. \tag{5.25}$$

Furthermore, (5.25) implies that for any $t \in [0, T_X]$,

(5.26)
$$\big|(\Phi_\omega + \eta(t), \partial_\omega \Phi_\omega)_{L^2}\big| \ge \big|(\Phi_\omega, \partial_\omega \Phi_\omega)_{L^2}\big| - \|\eta(t)\|_{L^2}\|\partial_\omega \Phi_\omega\|_{L^2} \ge \frac{1}{2}\big|(\Phi_\omega, \partial_\omega \Phi_\omega)_{L^2}\big|.$$

We see from Hölder's inequality and $N_\omega(\eta) = O(|\eta|^{\min\{2,p\}})$ (see (4.11)) that for any function $u \in L^1(\mathbb{R}^d) \cap L^\infty(\mathbb{R}^d)$,

$$\begin{aligned} \big|(N_\omega(\eta(t)), u)_{L^2_{real}}\big| &\le \|N_\omega(\eta(t))\|_{L^{\frac{p+1}{\min\{2,p\}}}} \|u\|_{L^{\frac{p+1}{\max\{1,p-1\}}}} \\ &\lesssim \|\eta(t)\|_{L^{p+1}}^{\min\{2,p\}} \|u\|_{L^1 \cap L^\infty}. \end{aligned} \tag{5.27}$$

We deduce from the equation (4.60) for $\theta(t)$, (5.26), (5.27) and (5.25) that for any $t \in [0, T_X]$,

$$\Big|\frac{d\theta}{dt}(t) - \omega\Big| \lesssim \frac{\|\eta(t)\|_{L^2}^2 + \|\eta(t)\|_{L^{p+1}}^{\min\{2,p\}}\|\partial_\omega \Phi_\omega\|_{L^1 \cap L^\infty}}{|(\Phi_\omega, \partial_\omega \Phi_\omega)_{L^2}|} \lesssim \|\eta(t)\|_{H^1}^{\min\{2,p\}}, \tag{5.28}$$

where the implicit constant depends on ω. Furthermore, it follows from (5.28), (5.27) and (5.25) that for any $t \in [0, T_X]$,

$$\begin{aligned} &\Big|\Omega\Big(\Big\{\frac{d\theta}{dt}(t) - \omega\Big\}\eta(t) - N_\omega(\eta(t)), f_2\Big)\Big| \\ &\le \Big|\frac{d\theta}{dt}(t) - \omega\Big| \big\|\eta(t)\big\|_{L^2}\|f_2\|_{L^2} + \|\eta(t)\|_{L^{p+1}}^{\min\{2,p\}}\|f_2\|_{L^2 \cap L^\infty} \\ &\lesssim \|\eta(t)\|_{H^1}^{\min\{2,p\}+1} + \|\eta(t)\|_{H^1}^{\min\{2,p\}} \lesssim \|\eta(t)\|_{H^1}^{\min\{2,p\}}, \end{aligned} \tag{5.29}$$

where the implicit constant depends on ω. Hence, we find form (5.29) and (5.19) that for any $t \in [0, T_X]$,

$$\left|\Omega\Big(\Big\{\frac{d\theta}{dt}(t) - \omega\Big\}\eta(t) - N_\omega(\eta(t)), f_2\Big)\right| \lesssim |\lambda_1(t)|^{\min\{2,\,p\}}, \tag{5.30}$$

where the implicit constant depends on ω. Similarly, we can verify that for any $t \in [0, T_X]$,

$$\left|\Big(\Big\{\frac{d\theta}{dt}(t) - \omega\Big\}\eta(t) - N_\omega(\eta(t)), \mathcal{U}_\pm\Big)_{L^2_{real}}\right| \lesssim |\lambda_1(t)|^{\min\{2,\,p\}}, \tag{5.31}$$

where the implicit constant depends on ω.

We shall show that

$$\lambda_1(0) \sim \lambda_+(0) \sim \mathfrak{s} R_0, \tag{5.32}$$

where the implicit constants depend on ω. We see from (5.9) and (5.11) that

$$\frac{d}{dt} d_\omega\big(\psi(t)\big)^2\Big|_{t=t_0=0} = 2R_0 \frac{d}{dt} d_\omega\big(\psi(t)\big)\Big|_{t=0} \ge 0, \tag{5.33}$$

which together with (5.20) yields

$$0 \le \mu^2 \lambda_1(0)\lambda_2(0) + \mu\lambda_1(0)\left|\Omega\Big(\Big\{\frac{d\theta}{dt}(0) - \omega\Big\}\eta(0) - N_\omega(\eta(0)), f_2\Big)\right|. \tag{5.34}$$

Combining (5.34) with (5.30), we obtain that

$$0 \le \mu^2 \operatorname{sgn}[\lambda_1(0)]|\lambda_1(0)|\lambda_2(0) + C\mu|\lambda_1(0)|^{\min\{3,p+1\}} \tag{5.35}$$

for some constant $C > 0$ depending on ω. Furthermore, it follows from (5.35) that

$$-C|\lambda_1(0)|^{\min\{2,p\}} \le \mu\mathfrak{s}\lambda_2(0). \tag{5.36}$$

Suppose here that $\mathfrak{s} = \operatorname{sgn}[\lambda_1(0)] = 1$. Then, we see from (5.22), (5.36) and (5.21) that

$$0 < \lambda_1(0) \lesssim \lambda_1(0) - \frac{C}{\mu}|\lambda_1(0)|^{\min\{2,p\}} \lesssim \lambda_1(0) + \lambda_2(0) = \lambda_+(0), \tag{5.37}$$

so that $\operatorname{sgn}[\lambda_+(0)] = 1$. Suppose next that $\operatorname{sgn}[\lambda_1(0)] = -1$. Then, (5.36) becomes $\mu\lambda_2(0) \le C|\lambda_1(0)|^{\min\{2,p\}}$. This together with (5.21) shows that

$$\lambda_+(0) = \lambda_1(0) + \lambda_2(0) \le \lambda_1(0) + \frac{C}{\mu}|\lambda_1(0)|^{\min\{2,p\}} \sim \lambda_1(0). \tag{5.38}$$

Thus, we conclude that

$$\operatorname{sgn}[\lambda_1(0)] = \operatorname{sgn}[\lambda_+(0)]. \tag{5.39}$$

Since $\lambda_1(0)$ and $\lambda_+(0)$ are independent of t, (5.39) together with (5.19) and $d_\omega(\psi(0)) = R_0$ implies (5.32).

Next, we shall prove (5.14) and (5.13). Since we have (5.19), it suffices for (5.14) to show that for any $t \in [0, T_X]$,

$$\lambda_1(t) \sim \lambda_+(t) \sim \mathfrak{s} R_0 e^{\mu t}. \tag{5.40}$$

Let $\alpha > 1$ be a constant satisfying $|\lambda_1(0)| \le \frac{1}{2}\alpha R_0$, and consider

$$T_\alpha := \sup\big\{\, T \in [0, T_X] \colon |\lambda_1(t)| \le \alpha R_0 e^{\mu t} \text{ for all } t \in [0, T] \,\big\}. \tag{5.41}$$

Then, we have $T_\alpha > 0$. We shall show that $T_\alpha = T_X$. Suppose for contradiction that $T_\alpha < T_X$. Then,

$$\alpha R_0 e^{\mu T_\alpha} = |\lambda_1(T_\alpha)| \lesssim \delta_X, \tag{5.42}$$

where we have used (5.21) to obtain the inequality. Using the equation (4.62) for λ_+, we have that

$$\begin{aligned} \frac{d}{dt}\big(e^{-\mu t}\lambda_+(t)\big) &= e^{-\mu t}\Big(\frac{d\lambda_+}{dt}(t) - \mu\lambda_+(t)\Big) \\ &= -e^{-\mu t}\big(\Big\{\frac{d\theta}{dt}(t)-\omega\Big\}\eta(t) - N_\omega(\eta(t)), \mathfrak{U}_-\big)_{L^2_{real}}. \end{aligned} \tag{5.43}$$

Furthermore, integrating the equation (5.43), and then using (5.31) and (5.21), we find that for any $t \in [0, T_X]$,

(5.44)

$$\begin{aligned} \big|\lambda_+(t) - e^{\mu t}\lambda_+(0)\big| &= \left| e^{\mu t}\int_0^t e^{-\mu s}\big(\Big\{\frac{d\theta}{dt}(s)-\omega\Big\}\eta(s) - N_\omega(\eta(s)), \mathfrak{U}_-\big)_{L^2_{real}} ds \right| \\ &\lesssim \int_0^t e^{\mu(t-s)}\big|\lambda_1(s)\big|^{\min\{2,p\}} ds, \end{aligned}$$

where the implicit constant depends on ω. Similarly, we have

$$\big|\lambda_-(t) - e^{-\mu t}\lambda_-(0)\big| \lesssim \int_0^t e^{-\mu(t-s)}\big|\lambda_1(s)\big|^{\min\{2,p\}} ds. \tag{5.45}$$

Then, we find from (5.44), (5.45), (5.32) and (5.21) that for any $t \in [0, T_X]$,

$$\begin{aligned} |\lambda_1(t)| &\le |\lambda_+(t)| + |\lambda_-(t)| \\ &\lesssim e^{\mu t}|\lambda_+(0)| + e^{-\mu t}|\lambda_-(0)| + \int_0^t e^{\mu(t-s)}\big|\lambda_1(s)\big|^{\min\{2,p\}} ds \\ &\lesssim e^{\mu t}R_0 + R_0 + e^{\mu t}\int_0^t e^{-\mu s}\big|\lambda_1(s)\big|^{\min\{2,p\}} ds. \end{aligned} \tag{5.46}$$

This together with the definition of T_α (see (5.41)) and (5.42) shows that

$$\begin{aligned} |\lambda_1(T_\alpha)| &\lesssim e^{\mu T_\alpha}R_0 + \frac{1}{\mu\min\{1,p-1\}}(\alpha R_0 e^{\mu T_\alpha})^{\min\{2,p\}} \\ &\le e^{\mu T_\alpha}R_0 + \frac{1}{\mu\min\{1,p-1\}}\delta_X^{\min\{1,p-1\}}\alpha R_0 e^{\mu T_\alpha}. \end{aligned} \tag{5.47}$$

Hence, if $\frac{1}{\mu\min\{1,p-1\}}\delta_X^{\min\{1,p-1\}} \ll 1$, then (5.47) implies that $|\lambda_1(T_\alpha)| \le \frac{1}{2}\alpha R_0 e^{\mu T_\alpha}$. However, this contradicts (5.42). Thus, we have proved that $T_\alpha = T_X$, and therefore for any $t \in [0, T_X]$,

$$|\lambda_1(t)| \lesssim R_0 e^{\mu t}. \tag{5.48}$$

We also see from (5.44), (5.45) and (5.48) that for any $t \in [0, T_X]$,

$$\big|\lambda_+(t) - e^{\mu t}\lambda_+(0)\big| \lesssim (R_0 e^{\mu t})^{\min\{2,p\}}, \tag{5.49}$$

$$\big|\lambda_-(t) - e^{-\mu t}\lambda_-(0)\big| \lesssim (R_0 e^{\mu t})^{\min\{2,p\}}. \tag{5.50}$$

Furthermore, we find from (5.32), (5.49) and (5.50) that (5.40) holds. Combining (5.19) and (5.40), we also obtain (5.13).

We shall prove (5.15). It follows from (5.40) that for any $t \in [0, T_X]$,

$$|\lambda_+(t)| + |\lambda_-(t)| \lesssim |\lambda_+(t)| + |\lambda_1(t)| \sim R_0 e^{\mu t}. \tag{5.51}$$

Furthermore, (5.50) together with (5.51) shows that

$$\bigl|\lambda_-(t)\bigr| \lesssim e^{-\mu t}\bigl|\lambda_-(0)\bigr| + (R_0 e^{\mu t})^{\min\{2,p\}} \lesssim R_0 + (R_0 e^{\mu t})^{\min\{2,p\}}. \tag{5.52}$$

In order to complete the proof of (5.15), we employ the "nonlinear energy projected onto $\mathfrak{U}_\pm$ plane":

$$E_{\{\mathfrak{U}_+,\mathfrak{U}_-\}}(t) := \mathcal{S}_\omega\bigl(\Phi_\omega + \lambda_+(t)\mathfrak{U}_+ + \lambda_-(t)\mathfrak{U}_-\bigr) - \mathcal{S}_\omega\bigl(\Phi_\omega\bigr). \tag{5.53}$$

The second order Taylor's expansion around Φ_ω together with $\mathcal{S}_\omega'(\Phi_\omega) = 0$, (4.3), $\mathcal{L}_\omega \mathfrak{U}_\pm = \pm i\mu \mathfrak{U}_\pm$ and (4.45) shows that

$$\begin{aligned}
E_{\{\mathfrak{U}_+,\mathfrak{U}_-\}}(t) &= \frac{1}{2}\bigl[\mathcal{S}_\omega''(\Phi_\omega)\{\lambda_+(t)\mathfrak{U}_+ + \lambda_-(t)\mathfrak{U}_-\}\bigr]\{\lambda_+(t)\mathfrak{U}_+ + \lambda_-(t)\mathfrak{U}_-\} \\
&\quad + O(\bigl\|\lambda_+(t)\mathfrak{U}_+ + \lambda_-(t)\mathfrak{U}_-\bigr\|_{H^1}^{\min\{3,p+1\}}) \\
&= \frac{1}{2}\langle \mathcal{L}_\omega\bigl(\lambda_+(t)\mathfrak{U}_+ + \lambda_-(t)\mathfrak{U}_-\bigr), \lambda_+(t)\mathfrak{U}_+ + \lambda_-(t)\mathfrak{U}_-\rangle_{H^{-1},H^1} \\
&\quad + O(\bigl\|\lambda_+(t)\mathfrak{U}_+ + \lambda_-(t)\mathfrak{U}_-\bigr\|_{H^1}^{\min\{3,p+1\}}) \\
&= \frac{1}{2}\bigl(i\mu\lambda_+(t)\mathfrak{U}_+ - i\mu\lambda_-(t)\mathfrak{U}_-, \lambda_+(t)\mathfrak{U}_+ + \lambda_-(t)\mathfrak{U}_-\bigr)_{L^2_{real}} \\
&\quad + O(\bigl\|\lambda_+(t)\mathfrak{U}_+ + \lambda_-(t)\mathfrak{U}_-\bigr\|_{H^1}^{\min\{3,p+1\}}) \\
&= -\mu\lambda_+(t)\lambda_-(t) + O(\bigl\|\lambda_+(t)\mathfrak{U}_+ + \lambda_-(t)\mathfrak{U}_-\bigr\|_{H^1}^{\min\{3,p+1\}}).
\end{aligned} \tag{5.54}$$

We find from (4.69), (5.8), (5.54) and Lemma 4.0.2 that

$$\begin{aligned}
&\mathcal{S}_\omega(\psi) - \mathcal{S}_\omega(\Phi_\omega) - E_{\{\mathfrak{U}_+,\mathfrak{U}_-\}}(t) \\
&= \frac{1}{2}\langle \mathcal{L}_\omega \Gamma(t), \Gamma(t)\rangle_{H^{-1},H^1} \\
&\quad + O(\|\eta(t)\|_{H^1}^{\min\{3,p+1\}}) + O(\bigl\|\lambda_+(t)\mathfrak{U}_+ + \lambda_-(t)\mathfrak{U}_-\bigr\|_{H^1}^{\min\{3,p+1\}}) \\
&\sim \|\Gamma(t)\|_{H^1}^2 + O(\|\eta(t)\|_{H^1}^{\min\{3,p+1\}}) + O(\bigl\|\lambda_+(t)\mathfrak{U}_+ + \lambda_-(t)\mathfrak{U}_-\bigr\|_{H^1}^{\min\{3,p+1\}}).
\end{aligned} \tag{5.55}$$

Moreover, it follows from $\mathcal{S}'_\omega(\Phi_\omega)=0$, (4.3), (4.64) and (5.31) that

$$
\begin{aligned}
&\left|\frac{d}{dt}E_{\{\mathfrak{U}_+,\mathfrak{U}_-\}}(t)\right| \\
&=\left|\mathcal{S}'_\omega(\Phi_\omega+\lambda_+(t)\mathfrak{U}_++\lambda_-(t)\mathfrak{U}_-)\Big\{\frac{d\lambda_+}{dt}(t)\mathfrak{U}_++\frac{d\lambda_-}{dt}(t)\mathfrak{U}_-\Big\}\right| \\
&\lesssim\left|\Big[\mathcal{S}''_\omega(\Phi_\omega)\big(\lambda_+(t)\mathfrak{U}_++\lambda_-(t)\mathfrak{U}_-\big)\Big]\Big\{\frac{d\lambda_+}{dt}(t)\mathfrak{U}_++\frac{d\lambda_-}{dt}(t)\mathfrak{U}_-\Big\}\right| \\
&\quad+\|\lambda_+(t)\mathfrak{U}_++\lambda_-(t)\mathfrak{U}_-\|_{H^1}^{\min\{2,p\}}\Big\|\frac{d\lambda_+}{dt}(t)\mathfrak{U}_++\frac{d\lambda_-}{dt}(t)\mathfrak{U}_-\Big\|_{H^1} \\
&=\left|\langle\mathcal{L}_\omega(\lambda_+(t)\mathfrak{U}_++\lambda_-(t)\mathfrak{U}_-),\,\frac{d\lambda_+}{dt}(t)\mathfrak{U}_++\frac{d\lambda_-}{dt}(t)\mathfrak{U}_-\rangle_{H^{-1},H^1}\right| \\
&\quad+\|\lambda_+(t)\mathfrak{U}_++\lambda_-(t)\mathfrak{U}_-\|_{H^1}^{\min\{2,p\}}\Big\|\frac{d\lambda_+}{dt}(t)\mathfrak{U}_++\frac{d\lambda_-}{dt}(t)\mathfrak{U}_-\Big\|_{H^1} \\
&\lesssim\mu(|\lambda_-(t)|+|\lambda_+(t)|)^{\min\{3,p+1\}} \\
&\quad+\big(|\lambda_+(t)|+|\lambda_-(t)|\big)^{\min\{2,p\}}\Big(\Big|\frac{d\lambda_+}{dt}(t)\Big|+\Big|\frac{d\lambda_-}{dt}(t)\Big|\Big).
\end{aligned}
\tag{5.56}
$$

Here, the equations (4.62) and (4.63) together with (5.28) and (5.27) show that for any $t\in[0,T_X]$,

$$
\begin{aligned}
&\Big|\frac{d\lambda_\pm}{dt}(t)\Big|\le\mu|\lambda_\pm(t)|+\Big|\frac{d\theta}{dt}(t)-\omega\Big|\|\eta(t)\|_{L^2}\|\mathfrak{U}_\pm\|_{L^2}+\big|\big(N_\omega(\eta(t)),\,\mathfrak{U}_\mp\big)_{L^2_{real}}\big| \\
&\lesssim|\lambda_+(t)|+|\lambda_-(t)|+\|\eta(t)\|_{H^1}^{\min\{3,p+1\}}+\|\eta(t)\|_{H^1}^{\min\{2,p\}} \\
&\lesssim|\lambda_+(t)|+|\lambda_-(t)|+\big\|\lambda_+(t)\mathfrak{U}_++\lambda_-(t)\mathfrak{U}_-+\Gamma(t)\big\|_{H^1}^{\min\{2,p\}} \\
&\lesssim|\lambda_+(t)|+|\lambda_-(t)|+\|\Gamma(t)\|_{H^1}^{\min\{2,p\}},
\end{aligned}
\tag{5.57}
$$

where the implicit constant depends on ω. Putting (5.56) and (5.57) together, we find that for any $t\in[0,T_X]$,

$$
\begin{aligned}
&\left|\frac{d}{dt}E_{\{\mathfrak{U}_+,\mathfrak{U}_-\}}(t)\right| \\
&\lesssim\|\Gamma(t)\|_{H^1}^{\min\{2,p\}}\big(|\lambda_+(t)|+|\lambda_-(t)|\big)+\big(|\lambda_+(t)|+|\lambda_-(t)|\big)^{\min\{3,p+1\}},
\end{aligned}
\tag{5.58}
$$

where the implicit constant depends on ω. Furthermore, we find from (5.55), the decomposition (4.66) and (5.58) that for any $t \in [0, T_X]$,

$$
\begin{aligned}
&\|\Gamma(t)\|_{H^1}^2 \\
&\lesssim \mathcal{S}_\omega(\psi) - \mathcal{S}_\omega(\Phi_\omega) - E_{\{\mathcal{U}_+,\mathcal{U}_-\}}(0) - \big(E_{\{\mathcal{U}_+,\mathcal{U}_-\}}(t) - E_{\{\mathcal{U}_+,\mathcal{U}_-\}}(0)\big) \\
&\quad + O(\|\eta(t)\|_{H^1}^{\min\{3,p+1\}}) + O(\|\lambda_+(t)\mathcal{U}_+ + \lambda_-(t)\mathcal{U}_-\|_{H^1}^{\min\{3,p\}}) \\
&\lesssim \|\Gamma(0)\|_{H^1}^2 + \int_0^t \left|\frac{d}{dt}E_{\{\mathcal{U}_+,\mathcal{U}_-\}}(t')\right| dt' \\
&\quad + \sup_{0\le t'\le t} \|\Gamma(t') + \lambda_+(t')\mathcal{U}_+ + \lambda_-(t')\mathcal{U}_-\|_{H^1}^{\min\{3,p+1\}} \\
&\quad + \sup_{0\le t'\le t} \big(|\lambda_+(t')| + |\lambda_-(t')|\big)^{\min\{3,p+1\}} \\
&\lesssim \|\Gamma(0)\|_{H^1}^2 + \int_0^t \|\Gamma(t')\|_{H^1}^{\min\{2,p\}} \big(|\lambda_+(t')| + |\lambda_-(t')|\big)\, dt' \\
&\quad + \int_0^t \big(|\lambda_+(t')| + |\lambda_-(t')|\big)^{\min\{3,p+1\}}\, dt' \\
&\quad + \sup_{0\le t'\le t} \|\Gamma(t')\|_{H^1}^{\min\{3,p+1\}} + \sup_{0\le t'\le t} \big(|\lambda_+(t')| + |\lambda_-(t')|\big)^{\min\{3,p+1\}}.
\end{aligned}
\tag{5.59}
$$

Suppose here that $p < 2$, so that $\min\{3, p+1\} = p+1$. Note that $p + 1 < \frac{2}{2-p}$. Then, it follows from (5.59), Young's inequality, (5.19), (5.51) and (5.40) that for any $t \in [0, T_X]$,

$$
\begin{aligned}
\sup_{0\le t'\le t} \|\Gamma(t')\|_{H^1}^2 &\lesssim \|\Gamma(0)\|_{H^1}^2 + \int_0^t \big(|\lambda_+(t')| + |\lambda_-(t')|\big)^{\frac{2}{2-p}}\, dt' \\
&\quad + \int_0^t \big(|\lambda_+(t')| + |\lambda_-(t')|\big)^{p+1}\, dt' \\
&\quad + \sup_{0\le t'\le t} \big\{|\lambda_+(t')| + |\lambda_-(t')|\big\}^{p+1} \\
&\lesssim |\lambda_1(0)|^2 + \int_0^t |\lambda_1(t')|^{p+1}\, dt' + \sup_{0\le t'\le t} |\lambda_1(t')|^{p+1} \\
&\lesssim R_0^2 + \int_0^t \big(R_0 e^{\mu t'}\big)^{p+1}\, dt + \sup_{0\le t'\le t} \big(R_0 e^{\mu t'}\big)^{p+1} \\
&\lesssim R_0^2 + (R_0 e^{\mu t})^{p+1}.
\end{aligned}
\tag{5.60}
$$

This together with (5.52) gives us the desired result (5.15) for $p < 2$. Similarly, we can prove (5.15) for $p \ge 2$.

We shall prove that there exists $T_* > 0$ such that $d_\omega(\psi(t))$ is strictly increasing on $[T_* R_0^{\min\{1,p-1\}}, T_X]$. It follows from (5.19), (5.20) and (5.30) that for any $t \in$

$[0, T_X]$,

$$|\lambda_1(t)|\frac{d}{dt}d_\omega(\psi(t)) \sim d_\omega(\psi(t))\frac{d}{dt}d_\omega(\psi(t)) = \frac{1}{2}\frac{d}{dt}d_\omega(\psi(t))^2 \\ \geq 2\mu^2|\lambda_1(t)|\mathfrak{s}\lambda_2(t) - C_1\mu|\lambda_1(t)|^{\min\{3,p+1\}}, \tag{5.61}$$

so that

$$\frac{d}{dt}d_\omega(\psi(t)) \gtrsim \mu^2\mathfrak{s}\lambda_2(t) - C_1\mu|\lambda_1(t)|^{\min\{2,p\}}, \tag{5.62}$$

where $C_1 > 0$ is constant depending only on d, p and ω. Here, multiplying the equation (4.90) by $\mathfrak{s}$, integrating the resulting equation, and using (5.31) and (5.40), we obtain that

$$\mathfrak{s}\lambda_2(t) = \mathfrak{s}\lambda_2(0) + \mu\int_0^t \mathfrak{s}\lambda_1(t')\,dt' \\ - \mathfrak{s}\frac{1}{2}\int_0^t (\Big\{\frac{d\theta}{dt}(t') - \omega\Big\}\eta(t') - N_\omega(\eta(t')), \mathcal{U}_+ + \mathcal{U}_-)_{L^2_{real}}\,dt' \\ \geq \mathfrak{s}\lambda_2(0) + C_2R_0(e^{\mu t} - 1) - C_3\int_0^t (R_0e^{\mu t'})^{\min\{2,p\}}\,dt', \tag{5.63}$$

where C_2 and C_3 are some positive constants depending only on d, p and ω. Here, we see from (5.36) and (5.32) that

$$\mu\mathfrak{s}\lambda_2(0) \gtrsim -|\lambda_1(0)|^{\min\{2,p\}} \sim -R_0^{\min\{2,p\}}, \tag{5.64}$$

where the implicit constants depend on ω. Moreover, it follows from (5.40) and (5.21) that for any $t \in [0, T_X]$,

$$|\lambda_1(t)| \sim R_0e^{\mu t} \lesssim \delta_X, \tag{5.65}$$

so that, taking δ_X sufficiently small dependently on ω, we have

$$C_3\int_0^t (R_0e^{\mu t'})^{\min\{2,p\}}\,dt' \leq \frac{C_2}{2}\mu\int_0^t R_0e^{\mu t'}\,dt' = \frac{C_2}{2}R_0(e^{\mu t} - 1). \tag{5.66}$$

Putting the estimates (5.63), (5.64) and (5.66) together, we find that for any $t \in [0, T_X]$,

$$\mathfrak{s}\lambda_2(t) \geq \frac{C_2}{2}R_0(e^{\mu t} - 1) - C_4R_0^{\min\{2,p\}} \geq \frac{C_2}{2}R_0\mu t - C_4R_0^{\min\{2,p\}} \tag{5.67}$$

for some constants $C_4 > 0$ depending only on d, p and ω. Hence, if we choose $T_* \geq \frac{4C_4}{\mu C_2}$, then for any $t \in [T_*R_0^{\min\{1,p-1\}}, T_X]$,

$$\mathfrak{s}\lambda_2(t) \geq T_*C_2\mu R_0^{\min\{2,p\}}. \tag{5.68}$$

Furthermore, it follows from (5.50), (5.40) and (5.51) that

$$|\lambda_2(t)| \geq |\lambda_1(t)| - |\lambda_-(t)| \\ \geq |\lambda_1(t)| - e^{-\mu t}\lambda_-(0) - C_5(e^{\mu t}R_0)^{\min\{2,p\}} \\ \geq c_1R_0e^{\mu t} - C_6R_0e^{-\mu t} - C_5(R_0e^{\mu t})^{\min\{2,p\}} \tag{5.69}$$

for some constants $c_1 \in (0,1)$, $C_5 > 0$ and $C_6 > 0$ depending only on d, p and ω. Hence, if $t \geq \frac{1}{2\mu}\log\left(\frac{4C_6}{c_1}\right)$, then we find from (5.69) and (5.40) that

$$|\lambda_2(t)| \geq \frac{c_1}{10} R_0 e^{\mu t} \sim |\lambda_1(t)|. \tag{5.70}$$

Furthermore, we see from (5.62), (5.70) and (5.65) that for any $t \geq \frac{1}{2\mu}\log\left(\frac{4C_6}{c_1}\right)$,

$$\frac{d}{dt} d_\omega(\psi(t)) > 0. \tag{5.71}$$

On the other hand, if $t \leq \frac{1}{2\mu}\log\left(\frac{4C_6}{c_1}\right)$, then we have

$$|\lambda_1(t)| \sim R_0 e^{\mu t} \lesssim R_0. \tag{5.72}$$

Hence, choosing T_* suitably, and using (5.62), (5.68) and (5.72), we conclude that for any $t \in [T_* R_0^{\min\{1,p-1\}}, T_X]$ with $t \leq \frac{1}{2\mu}\log\left(\frac{4C_6}{c_1}\right)$,

$$\frac{d}{dt} d_\omega(\psi(t)) \geq \frac{\mu^3 C_2 T_*}{2} R_0^{\min\{2,p\}} - C_1 |\lambda_1(t)|^{\min\{2,p\}} > 0. \tag{5.73}$$

Putting (5.71) and (5.73) together, we obtain the desired result.

We shall prove (5.17). It follows from $d_\omega(\psi(0)) = R_0$ and the fundamental theorem of calculus that

$$\left| d_\omega(\psi(t)) - R_0 \right| \leq \int_0^t \left| \frac{d}{ds} d_\omega(\psi(s)) \right| ds. \tag{5.74}$$

Here, it follows from (5.19), (5.20), (5.30) and (5.40) that

$$\begin{aligned} |\lambda_1(t)| \left| \frac{d}{dt} d_\omega(\psi(t)) \right| &\sim \left| 2 d_\omega(\psi(t)) \frac{d}{dt} d_\omega(\psi(t)) \right| = \left| \frac{d}{dt} d_\omega(\psi(t))^2 \right| \\ &\lesssim |\lambda_1(t)||\lambda_2(t)| + |\lambda_1(t)|^{\min\{3,p+1\}} \\ &= |\lambda_1(t)||\lambda_+(t) - \lambda_1(t)| + |\lambda_1(t)|^{\min\{3,p+1\}} \lesssim |\lambda_1(t)|^2, \end{aligned} \tag{5.75}$$

where the implicit constants depend on ω. Furthermore, this together with (5.40) gives us that for any $t \in [0, T_* R_0^{\min\{1,p-1\}}]$

$$\left| \frac{d}{dt} d_\omega(\psi(t)) \right| \lesssim R_0 e^{\mu t}. \tag{5.76}$$

Putting (5.74) and (5.76) together, we find that for any $t \in [0, T_* R_0^{\min\{1,p-1\}}]$,

$$\left| d_\omega(\psi(t)) - R_0 \right| \lesssim \int_0^t R_0 e^{\mu s}\, ds \lesssim T_* R_0^{\min\{2,p\}} e^{\mu T_* R_0^{\min\{1,p-1\}}} \lesssim R_0^{\min\{2,p\}}, \tag{5.77}$$

where the implicit constants depend on ω. Thus, we have proved (5.17).

Finally, we shall prove (5.16). Taylor's expansion of $\mathcal{K}$ around Φ_ω together with $\mathcal{K}(\Phi_\omega) = 0$ and (4.151) shows that

$$\begin{aligned} \mathcal{K}(\psi(t)) &= \mathcal{K}(\Phi_\omega + \eta(t)) = \mathcal{K}'(\Phi_\omega)\eta(t) + O(\|\eta(t)\|_{H^1}^2) \\ &= -2\mu s_p \lambda_1(t)(\Phi_\omega, f_2)_{L^2} - 2(1-s_p)(2^*-2)\lambda_1(t)(\Phi_\omega^{2^*-1}, f_1)_{L^2} \\ &\quad - s_p(p-1)(\Phi_\omega^p, \Gamma(t))_{L^2_{real}} - (2^*-2)(\Phi_\omega^{2^*-1}, \Gamma(t))_{L^2_{real}} \\ &\quad + \omega\mathcal{M}(\eta(t)) + O(\|\eta(t)\|_{H^1}^2). \end{aligned} \tag{5.78}$$

Furthermore, multiplying the both sides above by $\mathfrak{s} = \operatorname{sgn}[\lambda_1]$, and using $(\Phi_\omega, f_2)_{L^2} < 0$ (see (4.121)), (5.7) and (5.19), we find that for any $t \in [0, T_X]$,

(5.79)
$$\begin{aligned}
\mathfrak{s}\mathcal{K}(\psi(t)) &\ge 2s_p\mu\big|(\Phi_\omega, f_2)_{L^2}\big||\lambda_1(t)| - 2(1-s_p)(2^*-2)\big|(\Phi_\omega^{2^*-1}, f_1)_{L^2}\big||\lambda_1(t)| \\
&\quad - (p-1)\|\Phi_\omega\|_{L^{p+1}}^p\|\Gamma(t)\|_{L^{p+1}} - (2^*-2)\|\Phi_\omega\|_{L^{2^*}}^{2^*-1}\|\Gamma(t)\|_{L^{2^*}} \\
&\quad - O(|\lambda_1(t)|^2) \\
&\ge s_p\mu\big|(\Phi_\omega, f_2)_{L^2}\big||\lambda_1(t)| \\
&\quad - (p-1)\|\Phi_\omega\|_{L^{p+1}}^p\|\Gamma(t)\|_{L^{p+1}} - (2^*-2)\|\Phi_\omega\|_{L^{2^*}}^{2^*-1}\|\Gamma(t)\|_{L^{2^*}} \\
&\quad - O(|\lambda_1(t)|^2).
\end{aligned}$$

Recall here that $|\lambda_1(t)| \lesssim \delta_X \ll 1$ for $t \in [0, T_X]$ (see (5.19)). Then, we conclude from (5.79), (5.40) and the proved result (5.15) that there exists a constant $C_* > 0$ depending only on d, p and ω such that

$$\mathfrak{s}\mathcal{K}(\psi(t)) \gtrsim R_0 e^{\mu t} - C_* R_0, \tag{5.80}$$

where the implicit constant depends on ω. Thus, we have completed the proof. □

CHAPTER 6

Modified distance function

In this section, we continue to study the decomposition of the form (4.1) with (4.48), (4.53) and (4.54). Our aim here is to introduce the "modified distance function" $\widetilde{d}_\omega$ (see Proposition 6.0.2). To this end, we introduce two distances between a function $u \in H^1(\mathbb{R}^d)$ and the orbit of Φ_ω:

$$\operatorname{dist}_{H^1}\big(u, \mathcal{O}(\Phi_\omega)\big) := \inf_{\theta\in\mathbb{R}} \|u - e^{i\theta}\Phi_\omega\|_{H^1}, \tag{6.1}$$

$$\operatorname{dist}_{L^2}\big(u, \mathcal{O}(\Phi_\omega)\big) := \inf_{\theta\in\mathbb{R}} \|u - e^{i\theta}\Phi_\omega\|_{L^2}. \tag{6.2}$$

LEMMA 6.0.1. *Assume $d \geq 3$, $1 + \frac{4}{d} < p < 2^* - 1$, and let ω_2 be the frequency given by Proposition* 4.0.1. *Then, for any $\omega \in (0, \omega_2)$, there exists $\gamma_1(\omega) > 0$ with the following property: let ψ be a solution to* (NLS) *satisfying $\mathcal{M}(\psi) = \mathcal{M}(\Phi_\omega)$. If $\operatorname{dist}_{H^1}\big(\psi(t), \mathcal{O}(\Phi_\omega)\big) \leq \gamma_1(\omega)$, then*

$$\|\eta(t)\|_E \lesssim \operatorname{dist}_{H^1}\big(\psi(t), \mathcal{O}(\Phi_\omega)\big) \leq \|\eta(t)\|_E, \tag{6.3}$$

and if $\operatorname{dist}_{L^2}\big(\psi(t), \mathcal{O}(\Phi_\omega)\big) \leq \gamma_1(\omega)$, then

$$\|\eta(t)\|_{L^2} \sim \operatorname{dist}_{L^2}\big(\psi(t), \mathcal{O}(\Phi_\omega)\big). \tag{6.4}$$

Here, η is the function appearing in the decomposition for ψ of the form (4.1) *with* (4.48), (4.53) *and* (4.54). *Moreover, the implicit constants in* (6.3) *and* (6.4) *depend on ω as well as d and p.*

PROOF OF LEMMA 6.0.1. Note first that Lemma 4.0.3 shows

$$\operatorname{dist}_{H^1}\big(\psi(t), \mathcal{O}(\Phi_\omega)\big) \leq \|\psi(t) - e^{i\theta(t)}\Phi_\omega\|_{H^1} = \|\eta(t)\|_{H^1} \lesssim \|\eta(t)\|_E. \tag{6.5}$$

Moreover, we can take a continuous function $\theta_0(t)$ of t such that

$$\|\psi(t) - e^{i\theta_0(t)}\Phi_\omega\|_{H^1} = \operatorname{dist}_{H^1}\big(\psi(t), \mathcal{O}(\Phi_\omega)\big). \tag{6.6}$$

Put $\eta_0(t) := \psi(t) - e^{i\theta_0(t)}\Phi_\omega$, so that $\|\eta_0(t)\|_{H^1} = \operatorname{dist}_{H^1}\big(\psi(t), \mathcal{O}(\Phi_\omega)\big)$. Then, it follows from $\Omega(\Phi_\omega, \partial_\omega\Phi_\omega) = (\Phi_\omega, i\partial_\omega\Phi_\omega)_{L^2_{real}} = 0$ and (4.53) that

$$\begin{aligned}\Omega(e^{-i\theta(t)}\eta_0(t), \partial_\omega\Phi_\omega) &= \Omega(e^{-i\theta(t)}\psi(t), \partial_\omega\Phi_\omega) - \Omega(e^{i(\theta_0(t)-\theta(t))}\Phi_\omega, \partial_\omega\Phi_\omega)\\ &= -\big(ie^{i(\theta_0(t)-\theta(t))}\Phi_\omega, \partial_\omega\Phi_\omega\big)_{L^2_{real}}\\ &= \sin\big(\theta_0(t) - \theta(t)\big)(\Phi_\omega, \partial_\omega\Phi_\omega)_{L^2_{real}}.\end{aligned} \tag{6.7}$$

Hence, we have

$$\begin{aligned}\big|\sin\big(\theta_0(t) - \theta(t)\big)\big|\big|(\Phi_\omega, \partial_\omega\Phi_\omega)_{L^2_{real}}\big| &\leq \|\eta_0(t)\|_{L^2}\|\partial_\omega\Phi_\omega\|_{L^2}\\ &\leq \operatorname{dist}_{H^1}\big(\psi(t), \mathcal{O}(\Phi_\omega)\big)\|\partial_\omega\Phi_\omega\|_{L^2}.\end{aligned} \tag{6.8}$$

We see from (6.8) and $\mathrm{dist}_{H^1}\big(\psi(t), \mathcal{O}(\Phi_\omega)\big) \ll 1$ that

$$\inf_{k\in\mathbb{Z}} \big|\theta_0(t) - \theta(t) + k\pi\big| \lesssim \mathrm{dist}_{H^1}\big(\psi(t), \mathcal{O}(\Phi_\omega)\big), \tag{6.9}$$

where the implicit constant depends on ω. Here, we can eliminate the case where k is odd in the above infimum. Indeed, it follows from the choice of $\theta(t)$ (see (4.54)) that

$$\begin{aligned} 0 &> (e^{-i\theta(t)}\psi(t), \partial_\omega\Phi_\omega)_{L^2_{real}} \\ &= \cos\big(\theta_0(t) - \theta(t)\big)(\Phi_\omega, \partial_\omega\Phi_\omega)_{L^2_{real}} + (e^{-i\theta(t)}\eta_0(t), \partial_\omega\Phi_\omega)_{L^2_{real}}, \end{aligned} \tag{6.10}$$

so that

$$-\cos\big(\theta_0(t) - \theta(t)\big)(\Phi_\omega, \partial_\omega\Phi_\omega)_{L^2_{real}} > (e^{-i\theta(t)}\eta_0(t), \partial_\omega\Phi_\omega)_{L^2_{real}}. \tag{6.11}$$

Thus, supposing for contradiction that $\theta_0(t) - \theta(t)$ lay near an odd multiple of π, we have

$$\frac{1}{2}(\Phi_\omega, \partial_\omega\Phi_\omega)_{L^2_{real}} > (e^{-i\theta(t)}\eta_0(t), \partial_\omega\Phi_\omega)_{L^2_{real}}. \tag{6.12}$$

However, this together with (iii) of Proposition 2.0.4 shows

$$1 \lesssim \frac{1}{2}\big|(\Phi_\omega, \partial_\omega\Phi_\omega)_{L^2_{real}}\big| < \|\eta_0(t)\|_{L^2}\|\partial_\omega\Phi_\omega\|_{L^2} \le \mathrm{dist}_{H^1}\big(\psi(t), \mathcal{O}(\Phi_\omega)\big)\|\partial_\omega\Phi_\omega\|_{L^2}, \tag{6.13}$$

which contradicts that $\mathrm{dist}_{H^1}\big(\psi(t), \mathcal{O}(\Phi_\omega)\big) \ll 1$. Since $\mathrm{dist}_{H^1}\big(\psi(t), \mathcal{O}(\Phi_\omega)\big) \ll 1$ implies that $\theta_0(t) - \theta(t)$ lies near an even multiple of π, we see from Lemma 4.0.3 and (6.9) that

$$\begin{aligned} \|\eta(t)\|_E &\lesssim \|\eta(t)\|_{H^1} = \|e^{-i\theta(t)}\psi(t) - \Phi_\omega\|_{H^1} \\ &\le \big|e^{i(\theta_0(t)-\theta(t))} - 1\big|\|\Phi_\omega\|_{H^1} + \|\eta_0(t)\|_{H^1} \\ &\lesssim \big|\cos\big(\theta_0(t) - \theta(t)\big) - 1\big| + \big|\sin\big(\theta_0(t) - \theta(t)\big)\big| + \|\eta_0(t)\|_{H^1} \\ &\lesssim \inf_{k\in\mathbb{Z}} |\theta_0(t) - \theta(t) - 2k\pi| + \|\eta_0(t)\|_{H^1} \\ &\lesssim \mathrm{dist}_{H^1}\big(\psi(t), \mathcal{O}(\Phi_\omega)\big). \end{aligned} \tag{6.14}$$

Putting (6.5) and (6.14) together, we obtain the desired result (6.3). The same argument as the above shows (6.4). □

Now, we recall that $H^1_\omega(\mathbb{R}^d)$ denotes the set $\{u \in H^1(\mathbb{R}^d) \colon \mathcal{M}(u) = \mathcal{M}(\Phi_\omega)\}$ (see (4.84)). The following proposition is an immediate consequence of Lemma 6.0.1:

PROPOSITION 6.0.2. *Assume $d \ge 3$, $1 + \frac{4}{d} < p < 2^* - 1$, and let ω_2 be the frequency given by Proposition 4.0.1. Then, for any $\omega \in (0, \omega_2)$, there exist a constant $\widetilde{\gamma}(\omega) \in (0, \delta_E(\omega))$ ($\delta_E(\omega)$ denotes the constant given by Proposition 4.0.4) and a continuous function $\widetilde{d}_\omega \colon H^1_\omega(\mathbb{R}^d) \to [0, \infty)$ such that:*

$$\widetilde{d}_\omega(u) \sim \mathrm{dist}_{H^1}\big(u, \mathcal{O}(\Phi_\omega)\big), \tag{6.15}$$

where the implicit constant depends only on d, p and ω; and if $\widetilde{d}_\omega(u) \le \widetilde{\gamma}(\omega)$, then

$$\widetilde{d}_\omega(u) = d_\omega(u), \tag{6.16}$$

where d_ω is the distance function defined by (4.85).

REMARK 6.0.1. *We can take the constant δ_X given by the ejection lemma (Lemma* 5.0.2*) so small that $\delta_X \le \widetilde{\gamma}(\omega)$. Hence, we may assume that $\delta_X < \widetilde{\gamma}(\omega)$ in what follows.*

LEMMA 6.0.3. *Assume $d \ge 3$ and $1+\frac{4}{d} < p < 2^*-1$, and let ω_2 be the frequency given by Proposition* 4.0.1. *Then, for any $\omega \in (0,\omega_2)$ and any $\delta > 0$, there exist $\varepsilon_0(\delta) > 0$ and $\kappa_1(\delta) > 0$ with the following properties: $\varepsilon_0(\delta)$ and $\kappa_1(\delta)$ are non-decreasing with respect to δ; and for any radial function $u \in H^1(\mathbb{R}^d)$ satisfying*

$$\mathcal{M}(u) = \mathcal{M}(\Phi_\omega), \tag{6.17}$$

$$\mathcal{S}_\omega(u) < m_\omega + \varepsilon_0(\delta), \tag{6.18}$$

$$\widetilde{d}_\omega(u) \ge \delta, \tag{6.19}$$

we have either

$$\mathcal{K}(u) \le -\kappa_1(\delta) \tag{6.20}$$

or

$$\mathcal{K}(u) \ge \min\left\{\kappa_1(\delta),\ \frac{1}{2}\|\nabla u\|_{L^2}^2\right\}. \tag{6.21}$$

Here, $\widetilde{d}_\omega$ denotes the function given by Proposition 6.0.2.

PROOF OF LEMMA 6.0.3. We see from Hölder's inequality that if $u \in H^1(\mathbb{R}^d)$ satisfies $\mathcal{M}(u) = \mathcal{M}(\Phi_\omega)$, then

$$\|u\|_{L^{p+1}}^{p+1} \le \|u\|_{L^2}^{p+1-\frac{d(p-1)}{2}} \|u\|_{L^{2^*}}^{\frac{d(p-1)}{2}} = \mathcal{M}(\Phi_\omega)^{\frac{p+1}{2}-\frac{d(p-1)}{4}} \|u\|_{L^{2^*}}^{\frac{d(p-1)}{2}}. \tag{6.22}$$

Since $2 < \frac{d(p-1)}{2} < 2^*$, we see from (6.22) and Sobolev's embedding that there exists $c_0 > 0$ depending only on d, p and ω such that if $u \in H^1(\mathbb{R}^d)$ satisfies $\mathcal{M}(u) = \mathcal{M}(\Phi_\omega)$ and $\|\nabla u\|_{L^2}^2 \le c_0$, then $\mathcal{K}(u) \ge \frac{1}{2}\|\nabla u\|_{L^2}^2$. Thus, for the desired result, it is sufficient to prove that for any $\delta > 0$, there exist $\varepsilon_0(\delta) > 0$ and $\kappa_0(\delta) > 0$ with the following property: for any radial function $u \in H^1(\mathbb{R}^d)$ satisfying (6.17) through (6.19) and $\|\nabla u\|_{L^2}^2 \ge c_0$, we have $|\mathcal{K}(u)| > \kappa_1(\delta)$. We prove this by contradiction argument. Hence, we suppose to the contrary that there exists $\delta_0 > 0$ with the following property: for any number $n \ge 1$, there exists a radial function $u_n \in H^1(\mathbb{R}^d)$ satisfying

$$\mathcal{M}(u_n) = \mathcal{M}(\Phi_\omega), \tag{6.23}$$

$$\mathcal{S}_\omega(u_n) < m_\omega + \frac{1}{n}, \tag{6.24}$$

$$\widetilde{d}_\omega(u_n) \ge \delta_0, \tag{6.25}$$

$$\inf_{n\ge 1} \|\nabla u_n\|_{L^2} \ge c_0, \tag{6.26}$$

$$|\mathcal{K}(u_n)| \le \frac{1}{n}. \tag{6.27}$$

Then, we see from (6.24) and (6.27) that

$$\frac{s_p}{d}\|\nabla u_n\|_{L^2}^2 \le \mathcal{I}_\omega(u_n) = \mathcal{S}_\omega(u_n) - \frac{2}{d(p-1)}\mathcal{K}(u_n) \le m_\omega + 2, \tag{6.28}$$

so that

$$\|u_n\|_{H^1}^2 \le 2\mathcal{M}(\Phi_\omega) + \frac{d}{s_p}(m_\omega + 2). \tag{6.29}$$

Hence, we can take a radial function $u_\infty \in H^1(\mathbb{R}^d)$ and a subsequence of $\{u_n\}$ (still denoted by the same symbol $\{u_n\}$) such that

$$\lim_{n\to\infty} u_n = u_\infty \qquad \text{weakly in } H^1(\mathbb{R}^d) \tag{6.30}$$

and for any $2 < q < 2^*$,

$$\lim_{n\to\infty} u_n = u_\infty \qquad \text{strongly in } L^q(\mathbb{R}^d). \tag{6.31}$$

We shall show that u_∞ is non-trivial by using the contradiction argument. Hence, suppose to the contrary that $u_\infty \equiv 0$. Then, we see from (6.31) that

$$\lim_{n\to\infty} \|u_n\|_{L^{p+1}} = 0. \tag{6.32}$$

Moreover, (6.27) together with (6.32) shows that, passing to some subsequence,

$$\lim_{n\to\infty} \|\nabla u_n\|_{L^2}^2 = \lim_{n\to\infty} \|u_n\|_{L^{2^*}}^{2^*}. \tag{6.33}$$

We see from the definition of σ (see (1.27)) and (6.33) that

$$\lim_{n\to\infty} \|\nabla u_n\|_{L^2}^2 \ge \sigma \lim_{n\to\infty} \|u_n\|_{L^{2^*}}^2 = \sigma \lim_{n\to\infty} \|\nabla u_n\|_{L^2}^{\frac{4}{2^*}}, \tag{6.34}$$

which together with (6.33) and (6.26) yields

$$\lim_{n\to\infty} \|u_n\|_{L^{2^*}}^{2^*} \ge \sigma^{\frac{d}{2}}. \tag{6.35}$$

Furthermore, we see from the definition of $\mathcal{J}_\omega$ (see (1.12) and (1.14)), (6.24), (6.27) and (6.35), we find that

$$\frac{1}{d}\sigma^{\frac{d}{2}} \le \lim_{n\to\infty} \frac{1}{d}\|u_n\|_{L^{2^*}}^{2^*} \le \lim_{n\to\infty} \mathcal{J}_\omega(u_n) \le \lim_{n\to\infty} \{\mathcal{S}_\omega(u_n) + \frac{1}{2}|\mathcal{K}(u_n)|\} \le m_\omega. \tag{6.36}$$

However, this contradicts (1.35) and therefore u_∞ is non-trivial.

Next, we shall prove $u_\infty = e^{i\theta_0}\Phi_\omega$ for some θ_0. We see from (6.24) and (6.27) that

$$\mathcal{I}_\omega(u_\infty) \le \lim_{n\to\infty} \mathcal{I}_\omega(u_n) \le \lim_{n\to\infty} \{\mathcal{S}_\omega(u_n) + \frac{2}{d(p-1)}|\mathcal{K}(u_n)|\} \le m_\omega. \tag{6.37}$$

Hence, from the point of view of (1.10), what we need to prove is that $\mathcal{K}(u_\infty) \le 0$. We prove this by contradiction. Hence, suppose to the contrary that $\mathcal{K}(u_\infty) > 0$. Then, the Brezis-Lieb lemma together with (6.27) shows

$$\lim_{n\to\infty} \mathcal{K}(u_n - u_\infty) = -\lim_{n\to\infty} \{\mathcal{K}(u_n) - \mathcal{K}(u_n - u_\infty)\} = -\mathcal{K}(u_\infty) < 0, \tag{6.38}$$

so that for any sufficiently large $n \in \mathbb{N}$,

$$\mathcal{K}(u_n - u_\infty) < 0. \tag{6.39}$$

Hence, we can take $\lambda_n < 1$ such that $\mathcal{K}(\lambda_n^{\frac{d}{2}}\{u_n(\lambda_n\cdot) - u_\infty(\lambda_n\cdot)\}) = 0$ (cf. Lemma 2.1 in [**3**]). Moreover, this implies that

(6.40)
$$\begin{aligned} m_\omega &\le \mathcal{I}_\omega\big(\lambda_n^{\frac{d}{2}}\{u_n(\lambda_n\cdot) - u_\infty(\lambda_n\cdot)\}\big) \le \mathcal{I}_\omega(u_n - u_\infty) \\ &= \mathcal{I}_\omega(u_n) - \mathcal{I}_\omega(u_\infty) + o_n(1) = \mathcal{S}_\omega(u_n) - \frac{2}{d(p-1)}\mathcal{K}(u_n) - \mathcal{I}_\omega(u_\infty) + o_n(1), \end{aligned}$$

which together with (6.24) and (6.27) shows

$$m_\omega \le m_\omega - \mathcal{I}_\omega(u_\infty). \tag{6.41}$$

Since $\mathcal{I}_\omega(u_\infty) > 0$, this is a contradiction. Hence, $\mathcal{K}(u_\infty) \le 0$. In particular, we find from (6.37) that u_∞ is a minimizer of the variational problem (1.10). Since any minimizer of $\mathcal{I}_\omega$ is also a ground state (see Proposition 1.2 in [**2**]), we conclude from the radial symmetry and non-triviality of u_∞, and Proposition 2.0.1 that $u_\infty = e^{i\theta_0}\Phi_\omega$ for some $\theta_0 \in \mathbb{R}$. In particular, $\mathcal{K}(u_\infty) = 0$ and $\mathcal{I}(u_\infty) = m_\omega$. However, the condition (6.25) prevents u_∞ from existing. Indeed, it follows from the weak convergence (6.30) and $\lim_{n\to\infty} \mathcal{I}_\omega(u_n) \le m_\omega = \mathcal{I}(u_\infty)$ that

$$\lim_{n\to\infty} \|u_n\|_{L^2} = \|u_\infty\|_{L^2}, \qquad \lim_{n\to\infty} \|\nabla u_n\|_{L^2} = \|\nabla u_\infty\|_{L^2}. \tag{6.42}$$

Hence, we find that

$$\lim_{n\to\infty} u_n = u_\infty \qquad \text{strongly in } H^1(\mathbb{R}^d). \tag{6.43}$$

Using the strong convergence (6.43), we have

$$\delta_0 < \inf_\theta \|u_n - e^{i\theta}\Phi_\omega\|_{H^1} \le \|u_n - e^{i\theta_0}\Phi_\omega\|_{H^1} = \|u_n - u_\infty\|_{H^1} = o_n(1). \tag{6.44}$$

However, this is a contradiction. Thus, we cannot take a sequence $\{u_n\}$ satisfying (6.23) through (6.27) and therefore the lemma holds. □

LEMMA 6.0.4. *Assume $d \ge 3$ and $1+\frac{4}{d} < p < 2^*-1$, and let ω_2 be the frequency given by Proposition* 4.0.1. *Then, for any $\omega \in (0, \omega_2)$ and any $\delta > 0$, there exists $\kappa_2(\delta) > 0$ such that for any radial function $u \in H^1(\mathbb{R}^d)$ satisfying $\mathcal{J}_\omega(u) \le m_\omega - \delta$, the following holds:*

$$\mathcal{K}(u) \ge \min\big\{\kappa_2(\delta),\ \frac{1}{2}\|\nabla u\|_{L^2}^2\big\}. \tag{6.45}$$

PROOF OF LEMMA 6.0.4. Let u be a function satisfying $\mathcal{J}_\omega(u) \le m_\omega - \delta$. Then, we see from (1.14) and (1.10) that

$$\|u\|_{L^2}^2 \le \frac{2}{\omega^2}\mathcal{J}_\omega(u) \le \frac{2}{\omega^2}m_\omega, \tag{6.46}$$

$$\mathcal{K}(u) > 0. \tag{6.47}$$

Furthermore, it follows from Hölder's inequality and (6.46) that

$$\|u\|_{L^{p+1}}^{p+1} \le \|u\|_{L^2}^{p+1-\frac{d(p-1)}{2}} \|u\|_{L^{2^*}}^{\frac{d(p-1)}{2}} \le \Big(\frac{2}{\omega^2}m_\omega\Big)^{\frac{p+1}{2}-\frac{d(p-1)}{4}} \|u\|_{L^{2^*}}^{\frac{d(p-1)}{2}}. \tag{6.48}$$

Hence, we find from (6.48), $2 < \frac{d(p-1)}{2} < 2^*$ and Sobolev's embedding that there exists $c_0 > 0$ depending only on d, p and ω such that if $u \in H^1(\mathbb{R}^d)$ satisfies $\mathcal{J}_\omega(u) \le m_\omega - \delta$ and $\|\nabla u\|_{L^2}^2 \le c_0$, then $\mathcal{K}(u) \ge \frac{1}{2}\|\nabla u\|_{L^2}^2$. Thus, for the desired result (6.45), it is sufficient to prove that for any $\delta > 0$, there exists $\kappa_2(\delta) > 0$ with the following property: if $u \in H^1(\mathbb{R}^d)$ is radial and satisfies that $\mathcal{J}_\omega(u) \le m_\omega - \delta$ and $\|\nabla u\|_{L^2}^2 \ge c_0$, then $\mathcal{K}(u) > \kappa_2(\delta)$. We prove this by contradiction. Hence, suppose to the contrary that there exists $\delta_0 > 0$ with the following property: for any $n \in \mathbb{N}$, there exists a radial function $u_n \in H^1(\mathbb{R}^d)$ such that

$$\mathcal{J}_\omega(u_n) \le m_\omega - \delta_0, \tag{6.49}$$

$$\inf_{n\ge 1} \|\nabla u_n\|_{L^2} \ge c_0, \tag{6.50}$$

$$0 < \mathcal{K}(u_n) \le \frac{1}{n}. \tag{6.51}$$

Furthermore, it follows from (1.12), (6.49) and (6.51) that

$$\mathcal{S}_\omega(u_n) \le \mathcal{J}_\omega(u_n) + \frac{n}{2} \le m_\omega + \frac{n}{2} - \delta_0. \tag{6.52}$$

Now, using (1.14), (6.49) and (6.51), we find that for any $n \ge 1$,

$$\begin{aligned} \|u\|_{L^2}^2 + \|\nabla u_n\|_{L^2}^2 &\le \mathcal{K}(u_n) + \|u\|_{L^2}^2 + \|u\|_{L^{p+1}}^{p+1} + \|u\|_{L^{2^*}}^{2^*} \\ &\lesssim 1 + \mathcal{J}_\omega(u_n) \le 1 + m_\omega. \end{aligned} \tag{6.53}$$

Hence, we can extract a subsequence of $\{u_n\}$ (still denoted by the same symbol $\{u_n\}$) and a radial function $u_\infty \in H^1(\mathbb{R}^d)$ such that

$$\lim_{n\to\infty} u_n = u_\infty \qquad \text{weakly in } H^1(\mathbb{R}^d) \tag{6.54}$$

and for any $2 < q < 2^*$,

$$\lim_{n\to\infty} u_n = u_\infty \qquad \text{strongly in } L^q(\mathbb{R}^d). \tag{6.55}$$

Then, we see from the same argument as the proof of Lemma 6.0.3 that u_∞ is non-trivial. Furthermore, the lower semi-continuity of the weak limit together with (6.54), (6.55) and (6.49) implies that

$$\mathcal{J}_\omega(u_\infty) \le \liminf_{n\to\infty} \mathcal{J}_\omega(u_n) < m_\omega. \tag{6.56}$$

Hence, we find from (1.10) and $u_\infty \not\equiv 0$ that $\mathcal{K}(u_\infty) > 0$. However, the same argument as the proof of Lemma 6.0.3 shows that $\mathcal{K}(u_\infty) \le 0$. This is a contradiction. Thus, we have completed the proof. □

CHAPTER 7

One-pass theorem

In this section, we derive the one-pass theorem for our equation (NLS) (cf. Theorem 4.1 in [**25**]). To this end, we need the following inequalities for radial functions:

LEMMA 7.0.1. *The following hold for all $M > 0$ and all radial functions g in $H^1(\mathbb{R}^d)$: if $d = 3$ and $3 \le q \le 5 = 2^* - 1$, then*

$$\int_{\mathbb{R}^3} \frac{M|x|}{(M+|x|)^2}|g(x)|^{q+1}\,dx \lesssim \frac{1}{M}\|g\|_{H^1}^{q-1}\Big\|\nabla\Big(\frac{M}{M+|x|}g\Big)\Big\|_{L^2}^2, \tag{7.1}$$

and if $d \ge 4$ and $1 + \frac{4}{d-1} \le q \le 2^* - 1$, then

$$\begin{aligned}\int_{\mathbb{R}^d} \frac{M|x|}{(M+|x|)^2}|g(x)|^{q+1}\,dx &\lesssim M^{-\frac{(d-2)q-d}{4}}\|\nabla g\|_{L^2}^{q-1}\Big\|\nabla\Big(\frac{M}{M+|x|}g\Big)\Big\|_{L^2}^2 \\ &\quad + M^{-\frac{(d-1)(q-1)-4}{4}}\|g\|_{H^1}^{q-1}\int_{\mathbb{R}^d}\frac{M}{|x|(M+|x|)^2}|g(x)|^2\,dx.\end{aligned} \tag{7.2}$$

Here, the implicit constants depend only on d and q.

REMARK 7.0.1. *We rely on the radial Sobolev inequalities to prove Lemma* 7.0.1, *which causes the restriction of p in the one-pass theorem (Theorem* 7.0.1*) and Theorem* 1.0.1.

PROOF OF LEMMA 7.0.1. First, we prove the inequality (7.1). We see from (C.1) and Sobolev's embedding that for any $3 \le q \le 5$,

$$\begin{aligned}\int_{\mathbb{R}^3} \frac{M|x|}{(M+|x|)^2}|g(x)|^{q+1}\,dx &\le \frac{1}{M}\sup_{x\in\mathbb{R}^3}|x|\Big|\frac{M}{M+|x|}g(x)\Big|^2\int_{\mathbb{R}^3}|g(x)|^{q-1}\,dx \\ &\lesssim \frac{1}{M}\|g\|_{H^1}^{q-1}\Big\|\nabla\Big(\frac{M}{M+|x|}g\Big)\Big\|_{L^2}^2.\end{aligned} \tag{7.3}$$

Next, we prove (7.2). We see from (C.1), Hölder's inequality and Hardy's inequality that for any $d \ge 4$ and $1+\frac{4}{d} \le q \le 2^*-1$ (hence $\frac{d}{d-2} \le q \le 3$),

$$\begin{aligned}
&\int_{|x|\le\sqrt{M}} \frac{M|x|}{(M+|x|)^2}|g(x)|^{q+1}\,dx \\
&\le \frac{1}{M}\Big\| |x|^{\frac{d-2}{2}} \frac{M}{M+|x|} g\Big\|_{L^\infty}^2 \int_{|x|\le\sqrt{M}} \frac{|g(x)|^{q-1}}{|x|^{d-3}}\,dx \\
&\lesssim \frac{1}{M}\Big\|\nabla\Big(\frac{M}{M+|x|}g\Big)\Big\|_{L^2}^2 \Big(\int_{|x|\le\sqrt{M}} |x|^{-\frac{2(d-2-q)}{3-q}}\,dx\Big)^{\frac{3-q}{2}} \Big(\int_{\mathbb{R}^d} \frac{|g|^2}{|x|^2}\,dx\Big)^{\frac{q-1}{2}} \\
&\lesssim M^{-\frac{(d-2)q-d}{4}} \|\nabla g\|_{L^2}^{q-1} \Big\|\nabla\Big(\frac{M}{M+|x|}g\Big)\Big\|_{L^2}^2.
\end{aligned} \tag{7.4}$$

On the other hand, we see from (C.2) that for any $1+\frac{4}{d-1} \le q \le 2^*-1$,

$$\begin{aligned}
&\int_{\sqrt{M}\le|x|} \frac{M|x|}{(M+|x|)^2}|g(x)|^{q+1}\,dx \\
&\le \big\| |x|^{d-1}|g|^2\big\|_{L^\infty}^{\frac{q-1}{2}} \int_{\sqrt{M}\le|x|} \frac{|x|^{-\frac{(d-1)(q-1)}{2}+2}M}{|x|(M+|x|)^2}|g(x)|^2\,dx \\
&\lesssim M^{-\frac{(d-1)(q-1)-4}{4}} \|g\|_{H^1}^{q-1} \int_{\mathbb{R}^d} \frac{M}{|x|(M+|x|)^2}|g(x)|^2\,dx.
\end{aligned} \tag{7.5}$$

Putting (7.4) and (7.5) together, we obtain the desired estimate (7.2). □

Next, we recall that: ω_2 denotes the frequency given by Proposition 4.0.1; $\omega(2^*) \in (0,\omega_2)$ is the frequency satisfying (5.7); and for a given $\omega \in (0,\omega(2^*))$, δ_X, A_*, B_* and C_* denote the constants given by the ejection lemma (Lemma 5.0.2). Moreover, for given $\omega \in (0,\omega_2)$ and $\delta > 0$, $\varepsilon_0(\delta)$ denotes the constant determined by Lemma 6.0.3. Here, we may assume that

$$\delta_X \ll \mu|(\Phi_\omega, f_2)_{L^2}|. \tag{7.6}$$

Furthermore, we define $\delta_S > 0$ as a constant satisfying

$$\delta_S \le \frac{A_*\delta_X}{2B_*C_*} \quad \text{and} \quad \delta_S < \delta_X. \tag{7.7}$$

Note that $\delta_S < \delta_X < \delta_0(\omega) < \delta_E(\omega)$ and $\delta_X < \widetilde{\gamma}(\omega)$, where $\delta_E(\omega)$, $\delta_0(\omega)$ and $\widetilde{\gamma}(\omega)$ are the constants given by Proposition 4.0.4, Lemma 5.0.1 and Proposition 6.0.2, respectively.

In the proof of the one-pass theorem (Theorem 7.0.1) below, we use the following sign function:

$$\operatorname{sign}[\alpha] = \begin{cases} 1 & \text{if } \alpha \ge 0, \\ -1 & \text{if } \alpha < 0. \end{cases} \tag{7.8}$$

In particular, $\operatorname{sign}[0] = 1$.

THEOREM 7.0.1 (One-pass theorem). *Assume that either $d = 3$ and $3 \le p < 5$, or $d \ge 4$ and $1+\frac{4}{d-1} < p < 2^*-1$. Then, there exists $\omega_* > 0$ such that for* *any $\omega \in (0, \omega_*)$, there exist positive constants ε_*, δ_* and R_* with the following properties:*

$$\sqrt{\varepsilon_*} \ll R_* \ll \delta_* \ll \delta_S, \quad \varepsilon_* \le \varepsilon_0(\delta_*), \tag{7.9}$$

and for any $\varepsilon \in (0, \varepsilon_*]$, any $R \in (\sqrt{2\varepsilon}, R_*)$ and any radial solution ψ to (NLS) *satisfying*

$$\mathcal{M}(\psi) = \mathcal{M}(\Phi_\omega), \tag{7.10}$$

$$\mathcal{S}_\omega(\psi) < m_\omega + \varepsilon, \tag{7.11}$$

$$\widetilde{d}_\omega(\psi(0)) < R, \tag{7.12}$$

we have either

(i) $\widetilde{d}_\omega(\psi(t)) < R + R^{\frac{\min\{3,p+1\}}{2}}$ *for all $t \in [0, T_{\max}]$; or*

(ii) *there exists $t_* > 0$ such that $\widetilde{d}_\omega(\psi(t)) \ge R + R^{\frac{\min\{3,p+1\}}{2}}$ for all $t \in [t_*, T_{\max}]$. Here, $T_{\max}$ denotes the maximal lifespan of ψ.*

Proof of Theorem 7.0.1. We prove the claim by contradiction. Hence, we suppose to the contrary that for any $0 < \omega_* \ll 1$, there exists $0 < \omega < \omega_*$ such that for any $\varepsilon_*, \delta_*, R_* > 0$ with $\sqrt{\varepsilon_*} \ll R_* \ll \delta_* \ll \delta_S$ and $\varepsilon_* < \varepsilon_0(\delta_*)$, there exist $\varepsilon \in (0, \varepsilon_*]$, $R \in (\sqrt{2\varepsilon}, R_*)$ and a radial solution ψ such that for these ω, ε, R and ψ, the conditions (7.10), (7.11) and (7.12) hold, while both (i) and (ii) fail. Here, we may assume that δ_* is so small that

$$\log\left(\frac{\delta_X}{\delta_*}\right) \le \frac{1}{M_0}\frac{\delta_X}{\delta_*} \tag{7.13}$$

for some positive constant M_0 to be specified later dependently only on d, p and ω.

The failure of (i) together with (7.9) and (7.12) shows that there exist times $t_2 > t_1 > 0$ such that

$$\widetilde{d}_\omega(\psi(t_1)) = R < \delta_*, \tag{7.14}$$

$$\widetilde{d}_\omega(\psi(t_2)) = R + R^{\frac{\min\{3,p+1\}}{2}} < \delta_* \ll \delta_S < \delta_X, \tag{7.15}$$

$$R < \widetilde{d}_\omega(\psi(t)) < R + R^{\frac{\min\{3,p+1\}}{2}} < \delta_* \ll \delta_S < \delta_X \quad \text{for all } t \in (t_1, t_2). \tag{7.16}$$

Moreover, the failure of (ii) shows that there exists $t_3 > t_2$ such that

$$R < \widetilde{d}_\omega(\psi(t_3)) < R + R^{\frac{\min\{3,p+1\}}{2}}, \tag{7.17}$$

$$\widetilde{d}_\omega(\psi(t_3)) < \widetilde{d}_\omega(\psi(t)) \quad \text{for all } t \in (t_2, t_3). \tag{7.18}$$

We see from (7.14) through (7.18) that

$$R = \widetilde{d}_\omega(\psi(t_1)) = \min_{t \in [t_1, t_3]} \widetilde{d}_\omega(\psi(t)). \tag{7.19}$$

Note here that the ejection lemma (Lemma 5.0.2) together with (7.15) shows that

$$t_1 + T_* R^{\min\{1,p-1\}} < t_2, \tag{7.20}$$

where T_* is the constant given by the ejection lemma. Put

$$t_2' := \inf\{t \in [t_1, t_3] \colon \widetilde{d}_\omega(\psi(t)) = \delta_X\}. \tag{7.21}$$

Then, (7.16) shows $t_2 < t_2'$. Moreover, we see from (6.16) and the ejection lemma that

$$A_* e^{\mu(t-t_1)} R \le d_\omega(\psi(t)) \le B_* e^{\mu(t-t_1)} R, \tag{7.22}$$

$$\|\eta(t)\|_{H^1} \sim |\lambda_1(t)| \sim e^{\mu(t-t_1)} R \sim d_\omega(\psi(t)), \tag{7.23}$$

$$\|\Gamma(t)\|_{H^1} \lesssim R + \{e^{\mu(t-t_1)} R\}^{\frac{\min\{3,p+1\}}{2}} \sim \widetilde{d}_\omega(\psi(t_1)) + |\lambda_1(t)|^{\frac{\min\{3,p+1\}}{2}}, \tag{7.24}$$

$$\operatorname{sign}[\lambda_1(t)] \mathcal{K}(\psi(t)) \gtrsim (e^{\mu(t-t_1)} - C_*) R \tag{7.25}$$

for all $t \in [t_1, t_2']$, and $\widetilde{d}_\omega(\psi(t))$ increases on $[t_1 + T_* R^{\min\{1,p-1\}}, t_2']$. In particular, $\widetilde{d}_\omega(\psi(t))$ must reach δ_X before t_3 and therefore

$$t_2 < t_2' < t_3. \tag{7.26}$$

Furthermore, we see from (7.16) that there exists $t_S \in (t_2, t_2')$ such that $\widetilde{d}_\omega(\psi(t_S)) = \delta_S$.

Now, note that (7.11) together with $\sqrt{2\varepsilon} < R$ and (7.19) shows that

$$\mathcal{S}_\omega(\psi) < m_\omega + \frac{1}{2}\widetilde{d}_\omega(\psi(t))^2 \quad \text{for all } t \in [t_1, t_3]. \tag{7.27}$$

Hence, we find from Lemma 4.0.5, (6.16) and (7.19) that

$$R \le d_\omega(\psi(t)) \sim |\lambda_1(t)| \quad \text{for all } t \in [t_1, t_3] \text{ with } \widetilde{d}_\omega(\psi(t)) \le \delta_E(\omega). \tag{7.28}$$

In particular, $\operatorname{sign}[\lambda_1(t)]$ is constant on the set $\{t \in [t_1, t_3] \colon \widetilde{d}_\omega(\psi(t)) \le \delta_E(\omega)\}$. Moreover, Lemma 6.0.3 together with (7.11) and $\varepsilon < \varepsilon_* < \varepsilon_0(\delta_*)$ shows that $\operatorname{sign}[\mathcal{K}(\psi(t))]$ is constant on the set $\{t \in [t_1, t_3] \colon \widetilde{d}_\omega(\psi(t)) \ge \delta_*\}$.

We shall show that for any $t \in [t_1, t_3]$ with $\delta_* \le \widetilde{d}_\omega(\psi(t)) \le \delta_E(\omega)$,

$$\operatorname{sign}[\lambda_1(t)] = \operatorname{sign}[\mathcal{K}(\psi(t))]. \tag{7.29}$$

Since both $\operatorname{sign}[\lambda_1(t)]$ and $\operatorname{sign}[\mathcal{K}(\psi(t))]$ are constant on the set $\{t \in [t_1, t_3] \colon \delta_* \le \widetilde{d}_\omega(\psi(t)) \le \delta_E(\omega)\}$, it suffices to show that $\operatorname{sign}[\lambda_1(t_2')] = \operatorname{sign}[\mathcal{K}(\psi(t_2'))]$. We see from (7.21), (7.22), (6.16) and (7.7) that

$$\begin{aligned} \delta_X \le B_* e^{\mu(t_2'-t_1)} R &\le B_* e^{\mu(t_2'-t_1)} \frac{e^{-\mu(t_S-t_1)}}{A_*} \widetilde{d}_\omega(\psi(t_S)) \\ &= \frac{B_*}{A_*} e^{\mu(t_2'-t_S)} \delta_S \le \frac{\delta_X}{2C_*} e^{\mu(t_2'-t_S)}. \end{aligned} \tag{7.30}$$

Dividing the both sides of (7.30) by δ_X, and taking the logarithm, we obtain

$$\frac{1}{\mu} \log C_* < t_2' - t_S. \tag{7.31}$$

Since $t_1 < t_S$, we find from (7.25) and (7.31) that $\operatorname{sign}[\lambda_1(t_2')] = \operatorname{sign}[\mathcal{K}(\psi(t_2'))]$. Hence, (7.29) holds.

We find from (7.29) that the following function $\mathfrak{S}\colon [t_1,t_3]\to\{1,-1\}$ is well-defined:

$$\mathfrak{S}(t) := \begin{cases} \operatorname{sign}[\lambda_1(t)] & \text{if} \quad \widetilde{d}_\omega(\psi(t)) \le \delta_E(\omega), \\ \operatorname{sign}[\mathcal{K}(\psi(t))] & \text{if} \quad \widetilde{d}_\omega(\psi(t)) \ge \delta_*. \end{cases} \tag{7.32}$$

Note that the function $\mathfrak{S}(t)$ is constant on $[t_1,t_3]$. Put $\mathfrak{s} := \mathfrak{S}(t) \in \{1,-1\}$. When $\mathfrak{s}=1$, we see that there exists $C(\omega)>0$ such that for any $t\in[t_1,t_3]$,

$$\sup_{t\in[t_1,t_3]} \|\psi(t)\|_{H^1} \le C(\omega). \tag{7.33}$$

Indeed, if $\widetilde{d}_\omega(\psi(\tau)) \ge \delta_S$ for some $\tau\in[t_1,t_3]$, then (7.32) shows $\inf_{t\in[t_1,t_3]}\mathcal{K}(\psi(t)) \ge 0$. Hence, we see from (1.11), (1.13) and the assumption (7.11) that for any $t\in[t_1,t_3]$,

$$\frac{\omega}{2}\|\psi(t)\|_{L^2}^2 + \frac{s_p}{d}\|\nabla\psi(t)\|_{L^2}^2 \le \mathcal{I}_\omega(\psi(t)) \le \mathcal{S}_\omega(\psi) \le m_\omega + 1. \tag{7.34}$$

On the other hand, if $\widetilde{d}_\omega(\psi(t)) \le \delta_S$ for all $t\in[t_1,t_3]$, then we see from Lemma 4.0.3, Lemma 4.0.5 and (6.16) that for any $t\in[t_1,t_3]$,

$$\|\psi(t)\|_{H^1}^2 \lesssim \|\Phi_\omega\|_{H^1}^2 + \|\eta(t)\|_{H^1}^2 \lesssim \|\Phi_\omega\|_{H^1}^2 + d_\omega(\psi(t))^2 \lesssim \|\Phi_\omega\|_{H^1}^2 + \delta_S. \tag{7.35}$$

Thus, we have verified that (7.33) holds.

Let us summarize information obtained so far: there exist $t_1' (= t_1 + T_* R^{\min\{1,p-1\}}) \in (t_1,t_2)$ and $t_2'\in(t_2,t_3)$ such that: $\widetilde{d}_\omega(\psi(t))$ increases on $[t_1',t_2']$; $\widetilde{d}_\omega(\psi(t_2')) = \delta_X$; and for any $t\in(t_1,t_2')$,

$$\delta_X \ge \widetilde{d}_\omega(\psi(t)) \sim \|\eta(t)\|_{H^1} \sim |\lambda_1(t)| \sim e^{\mu(t-t_1)}R, \tag{7.36}$$

$$\|\Gamma(t)\|_{H^1} \lesssim \widetilde{d}_\omega(\psi(t_1)) + |\lambda_1(t)|^{\frac{\min\{3,p+1\}}{2}}, \tag{7.37}$$

$$\mathfrak{s}\mathcal{K}(\psi(t)) \gtrsim (e^{\mu(t-t_1)} - C_*)R. \tag{7.38}$$

Arguing from t_3 backward in time, we are also able to obtain a time interval $(t_2'',t_3)\subset(t_2',t_3)$ such that: $\widetilde{d}_\omega(\psi(t_2'')) = \delta_X$; for any $t\in(t_2'',t_3)$,

$$\delta_X \ge \widetilde{d}_\omega(\psi(t)) \sim \|\eta(t)\|_{H^1} \sim |\lambda_1(t)| \sim e^{\mu(t_3-t)}\widetilde{d}_\omega(\psi(t_3)), \tag{7.39}$$

$$\|\Gamma(t)\|_{H^1} \lesssim \widetilde{d}_\omega(\psi(t_3)) + |\lambda_1(t)|^{\frac{\min\{3,p+1\}}{2}}, \tag{7.40}$$

$$\mathfrak{s}\mathcal{K}(\psi(t)) \gtrsim (e^{\mu(t_3-t)} - C_*)\widetilde{d}_\omega(\psi(t_3)), \tag{7.41}$$

and $\widetilde{d}_\omega(\psi(t))$ decreases at least in the region $\{t\in[t_2'',t_3]\colon \widetilde{d}_\omega(\psi(t)) \ge 2\widetilde{d}_\omega(\psi(t_3))\}$ (cf. (5.17) in the ejection lemma (Lemma 5.0.2)).

Suppose here that there exists a time $\tau\in(t_2',t_2'')$ such that $\widetilde{d}_\omega(\psi(\tau))$ is a local minimum and $\widetilde{d}_\omega(\psi(\tau)) < \delta_*$. Then, we can apply the ejection lemma (Lemma 5.0.2) from τ both forward and backward in time to obtain an open interval $I_\tau\subset(t_2',t_2'')$ such that

$$\widetilde{d}_\omega(\psi(\inf I_\tau)) = \widetilde{d}_\omega(\psi(\sup I_\tau)) = \delta_X, \tag{7.42}$$

and for any $t \in I_\tau$,

$$\delta_X \geq \widetilde{d}_\omega(\psi(t)) \sim \|\eta(t)\|_{H^1} \sim |\lambda_1(t)| \sim e^{\mu|t-\tau|}\widetilde{d}_\omega(\psi(\tau)), \tag{7.43}$$

$$\|\Gamma(t)\|_{H^1} \lesssim \widetilde{d}_\omega(\psi(\tau)) + |\lambda_1(t)|^{\frac{\min\{3,p+1\}}{2}}, \tag{7.44}$$

$$\mathfrak{s}\mathcal{K}(\psi(t)) \gtrsim (e^{\mu|t-\tau|} - C_*)\widetilde{d}_\omega(\psi(\tau)), \tag{7.45}$$

and $\widetilde{d}_\omega(\psi(t))$ is monotone in the region $\widetilde{d}_\omega(\psi(t)) \geq 2\widetilde{d}_\omega(\psi(\tau))$. Note that for any distinct local minimum points τ_1 and τ_2 of $\widetilde{d}_\omega(\psi(t))$ in (t_2', t_2''), the monotonicity away from them implies that the intervals I_{τ_1} and I_{τ_2} are either disjoint or identical.

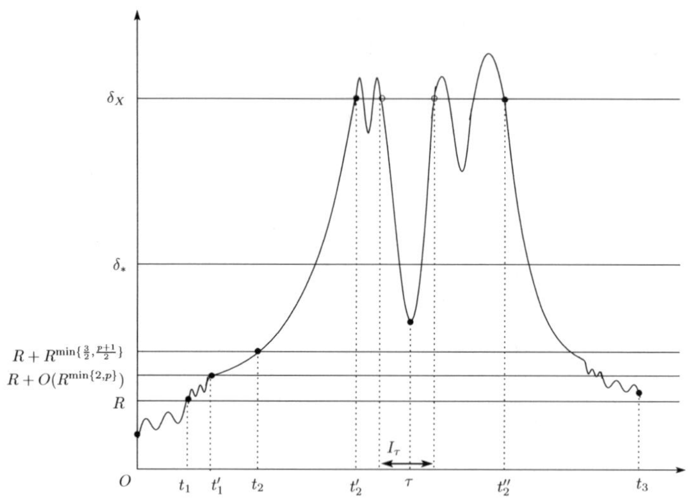

FIGURE 1. Behavior of $\widetilde{d}_\omega(\psi(t))$ in the case where there exists a local minimum point $\tau \in (t_2', t_2'')$ such that $\widetilde{d}_\omega(\psi(\tau)) < \delta_*$.

We find from the above observation that there exist a number $n \geq 2$ and disjoint open subintervals $I_1, \ldots, I_n$ of $[t_1, t_3]$ with the following properties[1]:

- $\inf I_1 = t_1$, $\sup I_n = t_3$, $\widetilde{d}_\omega(\psi(\sup I_1)) = \widetilde{d}_\omega(\psi(\inf I_n)) = \delta_X$, and

$$\widetilde{d}_\omega(\psi(\inf I_j)) = \widetilde{d}_\omega(\psi(\sup I_j)) = \delta_X \quad \text{except for } j = 1, n. \tag{7.46}$$

- For each $1 \leq j \leq n$, there exists $\tau_j \in I_j$ such that

$$\widetilde{d}_\omega(\psi(\tau_j)) < \delta_*, \tag{7.47}$$

and for any $t \in I_j$,

$$\delta_X \geq \widetilde{d}_\omega(\psi(t)) \sim \|\eta(t)\|_{H^1} \sim |\lambda_1(t)| \sim e^{\mu|t-\tau_j|}\widetilde{d}_\omega(\psi(\tau_j)), \tag{7.48}$$

$$\|\Gamma(t)\|_{H^1} \lesssim \widetilde{d}_\omega(\psi(\tau_j)) + |\lambda_1(t)|^{\frac{\min\{3,p+1\}}{2}}, \tag{7.49}$$

$$\mathfrak{s}\mathcal{K}(\psi(t)) \gtrsim (e^{\mu|t-\tau_j|} - C_*)\widetilde{d}_\omega(\psi(\tau_j)). \tag{7.50}$$

[1] Note that $I_1 = (t_1, t_2')$. Moreover, if $\widetilde{d}_\omega(\psi(t)) \geq \delta_*$ on $[t_2', t_2'']$, then $n = 2$ and $I_2 = (t_2'', t_3)$.

• If t satisfies $\sup I_{j-1} < t < \inf I_j$ for some $2 \le j \le n$, then

$$\widetilde{d}_\omega(\psi(t)) > \delta_*. \tag{7.51}$$

Put $I' := [t_1, t_3] \setminus \bigcup_{j=1}^n I_j$. Then, for any $t \in I'$, we have $\widetilde{d}_\omega(\psi(t)) > \delta_*$ (see (7.51)), which together with Lemma 6.0.3 gives us that for any $t \in I'$,

$$\begin{cases} \mathcal{K}(\psi(t)) \ge \min\{\kappa_1(\delta_*), \frac{1}{2}\|\nabla\psi(t)\|_{L^2}^2\} & \text{if} \quad \mathfrak{s} = 1, \\ \mathcal{K}(\psi(t)) \le -\kappa_1(\delta_*) & \text{if} \quad \mathfrak{s} = -1. \end{cases} \tag{7.52}$$

We can find from elementary calculations, (7.48), $\widetilde{d}_\omega(\psi(\tau_j)) \le \delta_*$ and (7.13) that for any $1 \le j \le n$,

$$\begin{aligned} \mu|I_j|\widetilde{d}_\omega(\psi(\tau_j)) &= \{\log e^{\mu|\sup I_j - \tau_j|} + \log e^{\mu|\inf I_j - \tau_j|}\}\widetilde{d}_\omega(\psi(\tau_j)) \\ &\lesssim \log\Big(\frac{\delta_X}{\widetilde{d}_\omega(\psi(\tau_j))}\Big)\widetilde{d}_\omega(\psi(\tau_j)) \le \frac{1}{M_0}\delta_X. \end{aligned} \tag{7.53}$$

We shall derive a contradiction in the case where $\mathfrak{s} = 1$. To this end, we use the following identity: for any $t \in I_{\max}(\psi)$ and any $M > 0$,

$$\begin{aligned} &\frac{d}{dt}\Im\int_{\mathbb{R}^d} \frac{M|x|}{M+|x|}\frac{x}{|x|}\cdot\nabla\psi(x,t)\overline{\psi(x,t)}\,dx \\ &= 2\mathcal{K}\Big(\frac{M}{M+|x|}\psi(t)\Big) \\ &\quad -\frac{p-1}{p+1}\int_{\mathbb{R}^d}\Big\{\frac{M^2}{(M+|x|)^2} + \frac{(d-1)M}{M+|x|} - \frac{dM^{p+1}}{(M+|x|)^{p+1}}\Big\}|\psi(t)|^{p+1}\,dx \\ &\quad -\frac{2}{d}\int_{\mathbb{R}^d}\Big\{\frac{M^2}{(M+|x|)^2} + \frac{(d-1)M}{M+|x|} - \frac{dM^{2^*}}{(M+|x|)^{2^*}}\Big\}|\psi(t)|^{2^*}\,dx \\ &\quad +\frac{1}{2}\int_{\mathbb{R}^d}\Big\{\frac{2M^2}{(M+|x|)^4} + \frac{(d-3)(d-1)M}{|x|(M+|x|)^2}\Big\}|\psi(t)|^2\,dx. \end{aligned} \tag{7.54}$$

Here, note that an elementary computation shows that for any $d \ge 3$ and any $1 \le q \le 2^* - 1$,

$$\begin{aligned} &\int_{\mathbb{R}^d}\Big\{\frac{M^2}{(M+|x|)^2} + \frac{(d-1)M}{M+|x|} - \frac{dM^{q+1}}{(M+|x|)^{q+1}}\Big\}|\psi(t)|^{q+1}\,dx \\ &= \int_{\mathbb{R}^d}\Big\{\frac{dM}{M+|x|}\Big(\frac{M}{M+|x|} - \frac{M^q}{(M+|x|)^q}\Big) + \frac{(d-1)M|x|}{(M+|x|)^2}\Big\}|\psi(t)|^{q+1}\,dx \\ &\lesssim \int_{\mathbb{R}^d}\frac{M|x|}{(M+|x|)^2}|\psi(t)|^{q+1}\,dx. \end{aligned} \tag{7.55}$$

We shall derive a contradiction in three steps:

Step 1. We shall derive an estimate for (7.54) on the region $\bigcup_{j=1}^n I_j$.

We see from Sobolev's embedding that for any $d \ge 3$, any $1 + \frac{4}{d} \le q \le 2^* - 1$, any $M > 0$ and any $t \in I_{\max}(\psi)$,

$$\begin{aligned} \int_{\mathbb{R}^d} \frac{M|x|}{(M+|x|)^2} |\psi(t)|^{q+1}\,dx &\lesssim \frac{1}{M}\int_{\mathbb{R}^d} |x||\Phi_\omega(x)|^{q+1}\,dx + \|\eta(t)\|_{L^{q+1}}^{q+1} \\ &\lesssim \frac{C(\omega)}{M} + \|\eta(t)\|_{H^1}^{q+1}, \end{aligned} \tag{7.56}$$

where $C(\omega)$ is some positive constant depending only on d, p and ω. Furthermore, it follows from (7.54), (7.55) and (7.56) that for any $d \ge 3$, any $1 + \frac{4}{d} \le p < 2^* - 1$, any $M > 0$ and any $t \in I_{\max}(\psi)$, then

$$\begin{aligned} &\frac{d}{dt}\Im \int_{\mathbb{R}^d} \frac{M|x|}{M+|x|}\frac{x}{|x|}\cdot \nabla\psi(x,t)\overline{\psi(x,t)}\,dx \\ &\ge 2\mathcal{K}\Big(\frac{M}{M+|x|}\psi(t)\Big) - \frac{C(\omega)}{M} - C\{\|\eta(t)\|_{H^1}^{p+1} + \|\eta(t)\|_{H^1}^{2^*}\} \end{aligned} \tag{7.57}$$

where $C(\omega)$ is some positive constant depending only on d, p and ω, and C is some positive constant depending only on d and p.

We consider the first two terms on the right-hand side of (7.57). Note first that it follows from (4.150) that for any $u \in H^1(\mathbb{R}^d)$,

$$\begin{aligned} \mathcal{K}'(\Phi_\omega)u = &-s_p\langle (p-1)\Phi_\omega^p, u\rangle_{H^{-1},H^1} - (2^*-2)\langle \Phi_\omega^{2^*-1}, u\rangle_{H^{-1},H^1} \\ &- 2\omega\langle \Phi_\omega, u\rangle_{H^{-1},H^1}. \end{aligned} \tag{7.58}$$

Furthermore, the first order Taylor's expansion of $\mathcal{K}$ around Φ_ω together with (7.58) shows that

(7.59)

$$\begin{aligned} &\mathcal{K}\Big(\frac{M}{M+|x|}\psi(t)\Big) = \mathcal{K}\Big(\Phi_\omega + \eta(t) - \frac{|x|}{M+|x|}\{\Phi_\omega + \eta(t)\}\Big) \\ &= -s_p\langle (p-1)\Phi_\omega^p, \eta(t)\rangle_{H^{-1},H^1} + s_p\langle (p-1)\Phi_\omega^p, \frac{|x|}{M+|x|}\{\Phi_\omega+\eta(t)\}\rangle_{H^{-1},H^1} \\ &\quad - (2^*-2)\langle \Phi_\omega^{2^*-1}, \eta(t)\rangle_{H^{-1},H^1} + (2^*-2)\langle \Phi_\omega^{2^*-1}, \frac{|x|}{M+|x|}\{\Phi_\omega+\eta(t)\}\rangle_{H^{-1},H^1} \\ &\quad - 2\omega\langle \Phi_\omega, \eta(t)\rangle_{H^{-1},H^1} + 2\omega\langle \Phi_\omega, \frac{|x|}{M+|x|}\{\Phi_\omega+\eta(t)\}\rangle_{H^{-1},H^1} \\ &\quad + O\Big(\Big\|\frac{M}{M+|x|}\eta(t) - \frac{|x|}{M+|x|}\Phi_\omega\Big\|_{H^1}^2\Big). \end{aligned}$$

Here, we see from the decomposition (4.108), (4.7), (4.109), (4.121) and $\mathfrak{s} = 1$ (hence $\lambda_1(t) \ge 0$ by (7.32)) that the first and third terms on the right-hand side of

(7.59) are rewritten as follows:

$$
\begin{aligned}
&- s_p\langle (p-1)\Phi_\omega^p, \eta(t)\rangle_{H^{-1},H^1} - (2^*-2)\langle \Phi_\omega^{2^*-1}, \eta(t)\rangle_{H^{-1},H^1} \\
&= -s_p\langle (p-1)\Phi_\omega^p, 2\lambda_1(t) f_1\rangle_{H^{-1},H^1} - s_p(p-1)\langle \Phi_\omega^p, \Gamma(t)\rangle_{H^{-1},H^1} \\
&\quad - (2^*-2)\langle \Phi_\omega^{2^*-1}, 2\lambda_1(t) f_1\rangle_{H^{-1},H^1} - (2^*-2)\langle \Phi_\omega^{2^*-1}, \Gamma(t)\rangle_{H^{-1},H^1} \\
&= 2s_p\lambda_1(t)\langle L_{\omega,+}\Phi_\omega, f_1\rangle_{H^{-1},H^1} - s_p(p-1)\langle \Phi_\omega^p, \Gamma(t)\rangle_{H^{-1},H^1} \\
&\quad - 2(1-s_p)(2^*-2)\lambda_1(t)\langle \Phi_\omega^{2^*-1}, f_1\rangle_{H^{-1},H^1} - (2^*-2)\langle \Phi_\omega^{2^*-1}, \Gamma(t)\rangle_{H^{-1},H^1} \\
&= 2s_p\mu|\lambda_1(t)||(\Phi_\omega, f_2)_{L^2}| - s_p(p-1)\langle \Phi_\omega^p, \Gamma(t)\rangle_{H^{-1},H^1} \\
&\quad - 2(1-s_p)(2^*-2)|\lambda_1(t)|\langle \Phi_\omega^{2^*-1}, f_1\rangle_{H^{-1},H^1} - (2^*-2)\langle \Phi_\omega^{2^*-1}, \Gamma(t)\rangle_{H^{-1},H^1}.
\end{aligned}
\tag{7.60}
$$

We also see from the decomposition (4.108), Lemma 4.0.7 and (4.149) that

$$
-2\omega\langle \Phi_\omega, \eta(t)\rangle_{H^{-1},H^1} = -2\omega\langle \Phi_\omega, \Gamma(t)\rangle_{H^{-1},H^1} = \omega\|\eta(t)\|_{L^2}^2. \tag{7.61}
$$

Moreover, it follows from Hölder's inequality, Sobolev's embedding and (2.9) that for any $p \le q \le 2^*-1$ and any sufficiently small $\omega > 0$,

$$
\left|(\Phi_\omega^q, \Gamma(t))_{L^2}\right| \le \|\Phi_\omega\|_{L^{q+1}}^q \|\Gamma(t)\|_{L^{q+1}} \lesssim \omega^{\frac{q}{q+1}(\frac{q-1}{p-1}-s_p)}\|\Gamma(t)\|_{H^1} \le \omega^{\frac{1-s_p}{2}}\|\Gamma(t)\|_{H^1}, \tag{7.62}
$$

where the implicit constant depends only on d and q; it follows from Hölder's inequality and Sobolev's embedding that for any $1 \le q \le 2^*-1$ and any sufficiently small $\omega > 0$,

$$
\begin{aligned}
&\langle \Phi_\omega^q, \frac{|x|}{M+|x|}\{\Phi_\omega + \eta(t)\}\rangle_{H^{-1},H^1} \ge -\int_{\mathbb{R}^d} \frac{|x|}{M+|x|}\Phi_\omega^q|\eta(t)|\,dx \\
&\ge -\frac{1}{M}\left(\int_{\mathbb{R}^d} |x|^{\frac{q+1}{q}}\Phi_\omega^{q+1}\,dx\right)^{\frac{q}{q+1}} \|\eta(t)\|_{L^{q+1}} \ge -\frac{1}{M}G(\omega)\|\eta(t)\|_{H^1},
\end{aligned}
\tag{7.63}
$$

where $G(\omega)$ is some positive constant depending only on d, q and ω; and it follows from Sobolev's embedding that

$$
\left\| \frac{M}{M+|x|}\eta(t) - \frac{|x|}{M+|x|}\Phi_\omega \right\|_{H^1}^2 \lesssim \|\eta(t)\|_{H^1}^2 + \frac{1}{M}G(\omega), \tag{7.64}
$$

where $G(\omega)$ is some positive constant depending only on d, q and ω.

Plugging (7.60) and (7.61) into (7.59), and then using Lemma 4.0.8 with $C = s_p - 2(1-s_p)(2^*-2)$, (7.62), (7.63) and (7.64), we find that if ω_* is sufficiently small (hence so is ω), then

$$\begin{aligned}
&\mathcal{K}\Big(\frac{M}{M+|x|}\psi(t)\Big)-\frac{C(\omega)}{M}\\
&=2s_p\mu|\lambda_1(t)||(\Phi_\omega,f_2)_{L^2}|-s_p(p-1)\langle\Phi_\omega^p,\Gamma(t)\rangle_{H^{-1},H^1}\\
&\quad-2(1-s_p)(2^*-2)|\lambda_1(t)|(\Phi_\omega^{2^*-1},f_1)_{L^2}-(2^*-2)\langle\Phi_\omega^{2^*-1},\Gamma(t)\rangle_{H^{-1},H^1}\\
&\quad+s_p\langle(p-1)\Phi_\omega^p,\frac{|x|}{M+|x|}\{\Phi_\omega+\eta(t)\}\rangle_{H^{-1},H^1}\\
&\quad+(2^*-2)\langle\Phi_\omega^{2^*-1},\frac{|x|}{M+|x|}\{\Phi_\omega+\eta(t)\}\rangle_{H^{-1},H^1}\\
&\quad+\omega\|\eta(t)\|_{L^2}^2+2\omega\langle\Phi_\omega,\frac{|x|}{M+|x|}\{\Phi_\omega+\eta(t)\}\rangle_{H^{-1},H^1}\\
&\quad+O\Big(\Big\|\frac{M}{M+|x|}\eta(t)-\frac{|x|}{M+|x|}\Phi_\omega\Big\|_{H^1}^2\Big)-\frac{C(\omega)}{M}\\
&\ge s_p\mu|\lambda_1(t)||(\Phi_\omega,f_2)_{L^2}|-C_1\omega^{\frac{1-s_p}{2}}\|\Gamma(t)\|_{H^1}-\frac{1}{M}G_1(\omega)\|\eta(t)\|_{H^1}\\
&\quad-C_2\|\eta(t)\|_{H^1}^2-\frac{1}{M}G_2(\omega),
\end{aligned}\tag{7.65}$$

where C_1 and C_2 are some positive constants depending only on d and p, and $G_1(\omega)$ and $G_2(\omega)$ are some positive constants depending only on d, q and ω. Recall here that (7.48), (7.49) and (7.6) give us that for any $t\in I_j$,

$$\|\eta(t)\|_{H^1}\sim|\lambda_1(t)|\sim e^{\mu|t-\tau_j|}\widetilde{d}_\omega(\psi(\tau_j)),\tag{7.66}$$

$$\|\eta(t)\|_{H^1}^2\lesssim\delta_X|\lambda_1(t)|\ll\mu|(\Phi_\omega,f_2)_{L^2}||\lambda_1(t)|,\tag{7.67}$$

$$\|\Gamma(t)\|_{H^1}\lesssim\widetilde{d}_\omega(\psi(\tau_j)).\tag{7.68}$$

Moreover, it follows from (7.19) that if $M\ge G_2(\omega)/R$, then

$$\frac{G_2(\omega)}{M}\le R\le\widetilde{d}_\omega(\psi(\tau_j)).\tag{7.69}$$

Hence, (7.65) together with (7.66) through (7.69) shows that for any sufficiently small $\omega>0$, any $M\ge G_2(\omega)/R$ and any $t\in I_j$,

$$\mathcal{K}\Big(\frac{M}{M+|x|}\psi(t)\Big)-\frac{C(\omega)}{M}\ge 2C_0^*(\omega)\{e^{\mu|t-\tau_j|}-C_1^*(\omega)\}\widetilde{d}_\omega(\psi(\tau_j)),\tag{7.70}$$

where $C_0^*(\omega)$ and $C_1^*(\omega)$ are some positive constant depending only on d, p and ω.

We find from (7.57), (7.70), $\|\eta(t)\|_{H^1}\lesssim\delta_X$ (see (7.48)) and (7.66) that if we take δ_X sufficiently small dependently only on d, p and ω in advance, then for any

$M \geq G_2(\omega)/R$ and any $t \in I_j$,

$$\begin{aligned}&\frac{d}{dt}\Im\int_{\mathbb{R}^d}\frac{M|x|}{M+|x|}\frac{x}{|x|}\cdot\nabla\psi(x,t)\overline{\psi(x,t)}\,dx\\&\geq 2C_0^*(\omega)\big\{e^{\mu|t-\tau_j|}-C_1^*(\omega)\big\}\widetilde{d}_\omega(\psi(\tau_j))-C\delta_X^{p-1}e^{\mu|t-\tau_j|}\widetilde{d}_\omega(\psi(\tau_j))\\&\geq C_0^*(\omega)\big\{e^{\mu|t-\tau_j|}-C_1^*(\omega)\big\}\widetilde{d}_\omega(\psi(\tau_j)).\end{aligned} \tag{7.71}$$

Step 2. Next, we consider the "variational region" I'. We have $\widetilde{d}_\omega(\psi) \geq \delta_*$ on I'. Moreover, it follows from (7.54), (7.55), Lemma 7.0.1 and (7.33) that if $d \geq 3$ and $3 \leq p < 5$, or $d \geq 4$ and $1+\frac{4}{d-1} < p < 2^*-1$, then for any sufficiently large $M > 0$ depending only on d, p and ω, we have

$$\begin{aligned}&\frac{d}{dt}\Im\int_{\mathbb{R}^d}\frac{M|x|}{M+|x|}\frac{x}{|x|}\cdot\nabla\psi(x,t)\overline{\psi(x,t)}\,dx\\&\geq 2\mathcal{K}\Big(\frac{M}{M+|x|}\psi(t)\Big)-o_M(1)\Big\|\nabla\Big(\frac{M}{M+|x|}\psi(t)\Big)\Big\|_{L^2}^2.\end{aligned} \tag{7.72}$$

Suppose here that $\|\nabla\psi(t)\|_{L^2} \leq \alpha$ for some $\alpha > 0$. Then, it follows from Hölder's inequality, $\frac{d(p-1)}{2} > 2$, Sobolev's embedding and the assumption (7.10) that

$$\begin{aligned}\Big\|\frac{M}{M+|x|}\psi(t)\Big\|_{L^{p+1}}^{p+1}&\leq\Big\|\frac{M}{M+|x|}\psi(t)\Big\|_{L^2}^{p+1-\frac{d(p-1)}{2}}\Big\|\frac{M}{M+|x|}\psi(t)\Big\|_{L^{2^*}}^{\frac{d(p-1)}{2}}\\&\leq\|\psi(t)\|_{L^2}^{p+1-\frac{d(p-1)}{2}}\|\psi(t)\|_{L^{2^*}}^{\frac{d(p-1)}{2}-2}\Big\|\frac{M}{M+|x|}\psi(t)\Big\|_{L^{2^*}}^2\\&\lesssim\mathcal{M}(\Phi_\omega)^{\frac{p+1}{2}-\frac{d(p-1)}{4}}\alpha^{\frac{d(p-1)}{2}-2}\Big\|\nabla\Big(\frac{M}{M+|x|}\psi(t)\Big)\Big\|_{L^2}^2.\end{aligned} \tag{7.73}$$

Similarly, we have

$$\Big\|\frac{M}{M+|x|}\psi(t)\Big\|_{L^{2^*}}^{2^*}\lesssim\alpha^{2^*-2}\Big\|\nabla\Big(\frac{M}{M+|x|}\psi(t)\Big)\Big\|_{L^2}^2. \tag{7.74}$$

Furthermore, we find from (7.73) and (7.74) that for a given $\omega > 0$ there exist $\alpha_0 > 0$ and $\kappa_0 > 0$ such that for any $t \in I_{\max}(\psi)$ with $\|\nabla\psi(t)\|_{L^2} \leq \alpha_0$,

$$\mathcal{K}\Big(\frac{M}{M+|x|}\psi(t)\Big)\geq\kappa_0\Big\|\nabla\Big(\frac{M}{M+|x|}\psi(t)\Big)\Big\|_{L^2}^2. \tag{7.75}$$

On the other hand, we see from (7.52) that for any $t \in I'$ with $\|\nabla\psi(t)\|_{L^2} > \alpha_0$,

$$\mathcal{K}(\psi(t))\geq\min\big\{\kappa_1(\delta_*),\ \frac{1}{2}\alpha_0^2\big\}. \tag{7.76}$$

Choosing $\varepsilon_* > 0$ so small that $\varepsilon_*^2 < \frac{1}{4}\min\{\kappa_1(\delta_*), \frac{1}{2}\alpha_0^2\}$, and using (7.11) and (7.76), we obtain that for any $t \in I'$ with $\|\nabla\psi(t)\|_{L^2} > \alpha_0$,

$$\mathcal{J}_\omega\Big(\frac{M}{M+|x|}\psi(t)\Big)\leq\mathcal{J}_\omega(\psi(t))=\mathcal{S}_\omega(\psi)-\frac{1}{2}\mathcal{K}(\psi(t))<m_\omega-\varepsilon_*^2, \tag{7.77}$$

which together with Lemma 6.0.4 gives us that

$$\mathcal{K}\Big(\frac{M}{M+|x|}\psi(t)\Big)\geq\min\{\kappa_2(\varepsilon_*^2),\ \frac{1}{2}\Big\|\nabla\Big(\frac{M}{M+|x|}\psi(t)\Big)\Big\|_{L^2}^2\}. \tag{7.78}$$

Putting (7.75) and (7.78) together, we find that for any $t \in I'$,

$$\mathcal{K}\Big(\frac{M}{M+|x|}\psi(t)\Big) \ge \min\{\kappa_2(\varepsilon_*^2),\ \kappa_0\Big\|\nabla\Big(\frac{M}{M+|x|}\psi(t)\Big)\Big\|_{L^2}^2\} \tag{7.79}$$

for some constant $\kappa_0 > 0$ depending only on d, p and ω. Note here that (7.33) shows

$$\Big\|\nabla\Big(\frac{M}{M+|x|}\psi(t)\Big)\Big\|_{L^2} \le \Big\|\frac{M}{(M+|x|)^2}\psi(t)\Big\|_{L^2} + \Big\|\frac{M}{M+|x|}\nabla\psi(t)\Big\|_{L^2} \lesssim C(\omega) \tag{7.80}$$

for all $t \in [t_1, t_3]$. Furthermore, we find from (7.72), (7.79) and (7.80) that if M satisfies $o_M(1) \ll \min\{\kappa_2(\varepsilon_*^2)/C(\omega)^2, \kappa_0\}$ and $t \in I'$, then

$$\frac{d}{dt}\Im\int_{\mathbb{R}^d}\frac{M|x|}{M+|x|}\frac{x}{|x|}\cdot\nabla\psi(x,t)\overline{\psi(x,t)}\,dx \ge 0. \tag{7.81}$$

Step 3. We finish the proof by deriving a contradiction.

We see from (7.71) and (7.81) that for any sufficiently small $\omega > 0$, there exists $L(\omega) \gg 1$ such that if $M \ge L(\omega)/R$, then

$$\begin{aligned}
&\Big[\Im\int_{\mathbb{R}^d}\frac{M|x|}{M+|x|}\frac{x}{|x|}\cdot\nabla\psi(x,t)\overline{\psi(x,t)}\,dx\Big]_{t_1}^{t_3}\\
&= \int_{t_1}^{t_3}\frac{d}{dt}\Im\int_{\mathbb{R}^d}\frac{M|x|}{M+|x|}\frac{x}{|x|}\cdot\nabla\psi(x,t)\overline{\psi(x,t)}\,dx\,dt\\
&\ge C_0^*(\omega)\sum_{j=1}^{n}\int_{I_j}\{e^{\mu|t-\tau_j|} - C_1^*(\omega)\}\widetilde{d}_\omega(\psi(\tau_j))\,dt\\
&= C_0^*(\omega)\sum_{j=1}^{n}\frac{1}{\mu}\Big\{e^{\mu|\sup I_j-\tau_j|} + e^{\mu|\inf I_j-\tau_j|} - 2\Big\}\widetilde{d}_\omega(\psi(\tau_j))\\
&\quad - C_0^*(\omega)C_1^*(\omega)\sum_{j=1}^{n}|I_j|\widetilde{d}_\omega(\psi(\tau_j)).
\end{aligned} \tag{7.82}$$

Here, it follows from (7.46) and (7.48) that for any $1 \le j \le n$,

$$\delta_X = \widetilde{d}_\omega(\psi(\sup I_j)) \sim e^{\mu|\sup I_j-\tau_j|}\widetilde{d}_\omega(\psi(\tau_j)) \quad \text{except for } j = n, \tag{7.83}$$

$$\delta_X = \widetilde{d}_\omega(\psi(\inf I_j)) \sim e^{\mu|\inf I_j-\tau_j|}\widetilde{d}_\omega(\psi(\tau_j)) \quad \text{except for } j = 1. \tag{7.84}$$

Moreover, we see from (7.53) and $\widetilde{d}_\omega(\psi(\tau_j)) \ge R$ (see (7.19)) that if M_0 is sufficiently large dependently only on d, p and ω,

$$\mu C_1^*(\omega)|I_j|\widetilde{d}_\omega(\psi(\tau_j)) \lesssim C_1^*(\omega)\frac{1}{M_0}\delta_X \ll \delta_X. \tag{7.85}$$

Thus, (7.82) together with (7.83), (7.84), $\widetilde{d}_\omega(\psi(\tau_j)) \le \delta_* \ll \delta_X$ and (7.85) shows that for any sufficiently large $M>0$ depending only on d, p, ω and R,

$$\begin{aligned}&\left[\Im \int_{\mathbb{R}^d} \frac{M|x|}{M+|x|}\frac{x}{|x|}\cdot \nabla\psi(x,t)\overline{\psi(x,t)}\right]_{t_1}^{t_3}\\ &\ge 2C^{**}(\omega)\sum_{j=1}^{n}\{\delta_X-\delta_*\} - C^{**}(\omega)\sum_{j=1}^{n}\delta_X \ge \frac{1}{2}C^{**}(\omega)n\delta_X \ge C^{**}(\omega)\delta_X\end{aligned} \tag{7.86}$$

for some positive constant $C^{**}(\omega)$ depending only on d, p and ω. Note here that

$$\left[\Im \int_{\mathbb{R}^d} \frac{M|x|}{M+|x|}\frac{x}{|x|}\cdot \nabla\Phi_\omega(x)\Phi_\omega(x)\,dx\right]_{t_1}^{t_3} = 0. \tag{7.87}$$

Moreover, we have

$$\left|\Im \int_{\mathbb{R}^d} \frac{M|x|}{M+|x|}\frac{x}{|x|}\cdot \Phi_\omega(x)\overline{\nabla\eta(x,t)}\,dx\right| \le \||x|\Phi_\omega\|_{L^2}\|\nabla\eta(t)\|_{L^2}. \tag{7.88}$$

Here, since Φ_ω decays exponentially, $\||x|\Phi_\omega\|_{L^2}$ is finite.

We see from the decomposition (4.1), (7.87), (7.88), Lemma 4.0.3, Lemma 4.0.5, $\|\eta(t_1)\|_{H^1} \sim \widetilde{d}_\omega(t_1) = R$ and $\|\eta(t_3)\|_{H^1} \sim \widetilde{d}_\omega(t_3) \lesssim R$ that if R_* is sufficiently small dependently on ω, and $M = L(\omega)/R$ for some sufficiently large constant $L(\omega)$ depending only on d, p and ω, then

$$\begin{aligned}&\left|\left[\Im \int_{\mathbb{R}^d} \frac{M|x|}{M+|x|}\frac{x}{|x|}\cdot \nabla\psi(x,t)\overline{\psi(x,t)}\,dx\right]_{t_1}^{t_3}\right|\\
&\le \left|\left[\Im \int_{\mathbb{R}^d} \frac{M|x|}{M+|x|}\frac{x}{|x|}\cdot \left(\nabla\eta(x,t)\Phi_\omega(x) + \nabla\Phi_\omega(x,t)\overline{\eta(x,t)}\right)\,dx\right]_{t_1}^{t_3}\right|\\
&\quad + \left|\left[\Im \int_{\mathbb{R}^d} \frac{M|x|}{M+|x|}\frac{x}{|x|}\cdot \nabla\eta(x,t)\overline{\eta(x,t)}\,dx\right]_{t_1}^{t_3}\right|\\
&\le \left|\left[\Im \int_{\mathbb{R}^d} \frac{M|x|}{M+|x|}\frac{x}{|x|}\cdot \nabla\eta(x,t)\Phi_\omega(x)\,dx\right]_{t_1}^{t_3}\right|\\
&\quad + \left|\left[\Im \int_{\mathbb{R}^d} \frac{M|x|}{M+|x|}\frac{x}{|x|}\cdot \Phi_\omega(x,t)\overline{\nabla\eta(x,t)}\,dx\right]_{t_1}^{t_3}\right|\\
&\quad + \left|\left[\Im \int_{\mathbb{R}^d} \left\{\frac{dM^2}{(M+|x|)^2} + \frac{(d-1)M|x|}{(M+|x|)^2}\right\}\Phi_\omega(x,t)\overline{\eta(x,t)}\,dx\right]_{t_1}^{t_3}\right|\\
&\quad + \left|\left[\Im \int_{\mathbb{R}^d} \frac{M|x|}{M+|x|}\frac{x}{|x|}\cdot \nabla\eta(x,t)\overline{\eta(x,t)}\,dx\right]_{t_1}^{t_3}\right|\\
&\lesssim \|\eta(t_1)\|_{H^1} + \|\eta(t_3)\|_{H^1} + M\|\eta(t_1)\|_{H^1}^2 + M\|\eta(t_3)\|_{H^1}^2 \ll C^{**}(\omega)\delta_X.\end{aligned} \tag{7.89}$$

This contradicts (7.86). Hence, the case $\mathfrak{s} = 1$ never happens.

We can deal with the case where $\mathfrak{s} = -1$ in a way similar to Section 4.1 of [**25**]. Hence, we have completed the proof. □

CHAPTER 8

Proof of Theorem 3.0.1

Our aim in this section is to prove Theorem 3.0.1. By the time reversality, it suffices to show that there are only three possibilities (scattering, blowup, trapping) forward in time. Throughout this section, we fix $\omega \in (0, \omega_*)$, where $\omega_* > 0$ is the frequency given by the one-pass theorem (Theorem 7.0.1).

We introduce several notation used in this section:

• We use $I_{\max}(\psi)$ to denote the maximal existence-interval of a solution ψ to (NLS), and set $T_{\max}(\psi) := \sup I_{\max}(\psi)$ and $T_{\min}(\psi) := \inf I_{\min}(\psi)$.

• Since the set A_ω^ε is invariant under the flow defined by (NLS) for all $\varepsilon > 0$ (see (3.1)), we use the convention $\psi \in A_\omega^\varepsilon$ to indicate that $\psi(t) \in A_\omega^\varepsilon$ for all $t \in I_{\max}(\psi)$.

• For given $R > 0$ and $\varepsilon > 0$, we define $S_{\omega,R}^\varepsilon$ by

$$S_{\omega,R}^\varepsilon := \left\{ \psi \colon \begin{array}{l} \psi \text{ is a radial solution to (NLS) such that } 0 \in I_{\max}(\psi), \\ \psi \in A_\omega^\varepsilon, \text{ and } \widetilde{d}_\omega(\psi(t)) \geq R \text{ for all } t \in [0, T_{\max}(\psi)) \end{array} \right\}. \tag{8.1}$$

Next, we give a fundamental fact in A_ω^ε:

LEMMA 8.0.1. *Assume $d \geq 3$ and $1 + \frac{4}{d} < p < 2^* - 1$. Let ψ be a solution to (NLS) and $\varepsilon > 0$. Furthermore, assume that $\psi \in A_\omega^\varepsilon$ and*

$$\inf_{t \in I_{\max}(\psi)} \|\nabla \psi(t)\|_{L^2}^2 \leq \mathcal{H}(\Phi_\omega). \tag{8.2}$$

Then, we have either

$$\sup_{t \in I_{\max}(\psi)} \mathcal{K}(\psi(t)) < 0, \tag{8.3}$$

or

$$\inf_{t \in I_{\max}(\psi)} \mathcal{K}(\psi(t)) > 0. \tag{8.4}$$

PROOF OF LEMMA 8.0.1. Note first that $\mathcal{M}(\psi) = \mathcal{M}(\Phi_\omega)$ and therefore ψ is non-trivial. We also see from the assumption (8.2) that there exists $t_0 \in I_{\max}(\psi)$ such that

$$\|\nabla \psi(t_0)\|_{L^2}^2 \leq \frac{3}{2}\mathcal{H}(\Phi_\omega). \tag{8.5}$$

Furthermore, we see from $\mathcal{M}(\psi) = \mathcal{M}(\Phi_\omega)$ and (8.5) that

$$\begin{aligned} \mathcal{S}_\omega(\psi) &= \omega\mathcal{M}(\psi) + \frac{1}{2}\|\nabla\psi(t_0)\|_{L^2}^2 - \frac{1}{p+1}\|\psi(t_0)\|_{L^{p+1}}^{p+1} - \frac{1}{2^*}\|\psi(t_0)\|_{L^{2^*}}^{2^*} \\ &\leq \omega\mathcal{M}(\Phi_\omega) + \frac{3}{4}\mathcal{H}(\Phi_\omega) = \mathcal{S}_\omega(\Phi_\omega) - \frac{1}{4}\mathcal{H}(\Phi_\omega) < m_\omega. \end{aligned} \tag{8.6}$$

Hence, the claim follows from (1.49) and Theorem 1.0.3. □

Now, let us recall that δ_X and δ_S denote the constants given by the ejection lemma (Lemma 5.0.2) and (7.7), respectively (see also Remark 6.0.1). Furthermore, the one-pass theorem shows that there exist constants $0 < \varepsilon_* \ll R_* \ll \delta_* (\ll \delta_S < \delta_X)$ such that: $\varepsilon_* \le \varepsilon_0(\delta_*)$ (see Lemma 6.0.3 for the definition of $\varepsilon_0(\delta_*)$); and for any $\varepsilon \in (0, \varepsilon_*)$, any $R \in (\sqrt{2\varepsilon}, R_*)$ and any solution ψ to (NLS) with $\psi \in A_\omega^\varepsilon$, the following alternative holds:

Case 1. There exists $t_0 \in I_{\max}(\psi)$ such that $\widetilde{d}_\omega(\psi(t_0)) < R$ and $\widetilde{d}_\omega(\psi(t)) < R + R^{\frac{\min\{3,p+1\}}{2}}$ for all $t \in [t_0, T_{\max}(\psi))$; or

Case 2. There exists $t_1 \in I_{\max}(\psi)$ such that for all $t_1 \le t < T_{\max}(\psi)$,

$$\widetilde{d}_\omega(\psi(t)) \ge R. \tag{8.7}$$

In Case 1, it follows from $R + R^{\frac{\min\{3,p+1\}}{2}} < \delta_S < \delta_X$, (6.16) in Proposition 6.0.2, Remark 6.0.1 and Lemma 5.0.1 that $T_{\max}(\psi) = \infty$. Hence, ψ is trapped by $\mathcal{O}(\Phi_\omega)$ forward in time.

Next, we consider Case 2. In this case, we have that $\psi(\cdot - t_1) \in S_{\omega,R}^\varepsilon$ and

$$\varepsilon \le \min\Big\{\frac{\widetilde{d}_\omega(\psi(t))}{2}, \varepsilon_0(\delta_*)\Big\} \tag{8.8}$$

for any $t \in [t_1, T_{\max}(\psi))$. In Section 8.0.1 and Section 8.0.2, we will prove that the solution ψ blows up in a finite time or scatters. To this end, we need some preparations. Let us begin with the following:

LEMMA 8.0.2. *Let $\varepsilon \in (0, \varepsilon_*)$, $R \in (\sqrt{2\varepsilon}, R_*)$, and let $\psi \in S_{\omega,R}^\varepsilon$. Then, for any $T \in (0, T_{\max}(\psi))$, there exists $\tau \in [T, T_{\max}(\psi))$ such that*

$$\widetilde{d}_\omega(\psi(\tau)) \ge \delta_X. \tag{8.9}$$

PROOF OF LEMMA 8.0.2. Suppose for contradiction that there exists $T_0 \in (0, T_{\max}(\psi))$ such that for any $t \in [T_0, T_{\max}(\psi))$,

$$d_\omega(\psi(t)) = \widetilde{d}_\omega(\psi(t)) < \delta_X, \tag{8.10}$$

where the equality follows from (6.16) in Proposition 6.0.2 and Remark 6.0.1. Then, we see from Lemma 5.0.1 that $T_{\max}(\psi) = \infty$. Let $L \gg \frac{1}{\mu}$, and consider a time $t_L \in [T_0, T_0 + L\log\frac{\delta_X}{\varepsilon_*}]$ such that $R_L := d_\omega(\psi(t_L))$ is the minimum of $d_\omega(\psi(t))$ over the interval $[T_0, T_0 + L\log\frac{\delta_X}{\varepsilon_*}]$. Note that $\delta_X > R_L \ge R > \varepsilon_*$. If $T_0 \le t_L \le T_0 + \frac{L}{2}\log\frac{\delta_X}{\varepsilon_*}$, then we see from the ejection lemma (Lemma 5.0.2) that

$$d_\omega\Big(\psi\Big(t_L + \frac{L}{2}\log\frac{\delta_X}{\varepsilon_*}\Big)\Big) \sim e^{\mu(\frac{L}{2}\log\frac{\delta_X}{\varepsilon_*})} R_L > \Big(\frac{\delta_X}{\varepsilon_*}\Big)^{\frac{\mu L}{2}} \varepsilon_* \gg \delta_X. \tag{8.11}$$

However, this contradicts (8.10). Similarly, when $t_L \ge T_0 + \frac{L}{2}\log\frac{\delta_X}{\varepsilon_*}$, applying the ejection lemma backward in time, we reach a contradiction. Thus, we have proved the claim. □

Next, we show that the solution ψ stays away from Φ_ω after some time.

LEMMA 8.0.3. *Let $\varepsilon \in (0, \varepsilon_*)$, $R \in (\sqrt{2\varepsilon}, R_*)$, and let $\psi \in S_{\omega,R}^\varepsilon$. Then, there exists a time $T_0 \in [0, T_{\max}(\psi))$ such that for any $t \in [T_0, T_{\max}(\psi))$,*

$$\widetilde{d}_\omega(\psi(t)) \ge R_*. \tag{8.12}$$

PROOF OF LEMMA 8.0.3. It suffices to consider the case where there exists a time $\tau_0 \in [0, T_{\max}(\psi))$ such that

$$R \le \widetilde{R}_* := \widetilde{d}_\omega\big(\psi(\tau_0)\big) < R_*. \tag{8.13}$$

Here, it follows from Lemma 8.0.2 that we may assume that

$$\frac{\sqrt{1+4R_*}-1}{2} \le \widetilde{R}_* < R_*. \tag{8.14}$$

Then, we see from the one-pass theorem (Theorem 7.0.1) that either

$$\sup_{t \ge \tau_0} \widetilde{d}_\omega\big(\psi(t)\big) \le \widetilde{R}_* + \widetilde{R}_*^{\frac{\min\{3,p+1\}}{2}}, \tag{8.15}$$

or there exists $T_0 \ge \tau_0$ such that

$$\inf_{t \ge T_0} \widetilde{d}_\omega\big(\psi(t)\big) \ge \widetilde{R}_* + \widetilde{R}_*^{\frac{\min\{3,p+1\}}{2}} \ge \widetilde{R}_* + \widetilde{R}_*^2 \ge R_*. \tag{8.16}$$

We find from Lemma 8.0.2 and $R_* \ll \delta_X$ that the former case (8.15) never happens. Thus, the latter case (8.16) only happens, and the proof is completed. □

Lastly, we introduce a sign function which will determine the scattering or blowup.

LEMMA 8.0.4. *Let $\varepsilon \in (0, \varepsilon_*)$ and $R \in (\sqrt{2\varepsilon}, R_*)$. Then, there exists a function $\mathfrak{S}\colon S^{\varepsilon}_{\omega,R} \to \{1,-1\}$ such that*

$$\mathfrak{S}(\psi) = \begin{cases} \operatorname{sign}[\lambda_1(t)] & \text{if } t \in [0, T_{\max}(\psi)) \text{ with } \widetilde{d}_\omega(\psi(t)) \le \delta_E(\omega), \\ \operatorname{sign}[\mathcal{K}(\psi(t))] & \text{if } t \in [0, T_{\max}(\psi)) \text{ with } \widetilde{d}_\omega(\psi(t)) \ge \delta_*. \end{cases} \tag{8.17}$$

In addition, there exists a positive constant $C(\omega)$ which is independent of ε and R and satisfies that

$$\sup\big\{\|\psi\|_{L^\infty_t H^1_x([0,T_{\max}(\psi)))} : \psi \in S^{\varepsilon}_{\omega,R},\ \mathfrak{S}(\psi) = 1\big\} \le C(\omega). \tag{8.18}$$

PROOF OF LEMMA 8.0.4. We introduced a sign function (7.32) in the proof of one-pass theorem (Theorem 7.0.1). Employing the same argument as (7.32) and Lemma 8.0.2, we can prove the existence of the desired sign function. Furthermore, we can prove (8.18) in the same way as (7.33). □

We divide $S^{\varepsilon}_{\omega,R}$ into two parts according to the sign of $\mathfrak{S}$:

$$S^{\varepsilon}_{\omega,R,\pm} := \big\{\psi \in S^{\varepsilon}_{\omega,R} \colon \mathfrak{S}(\psi) = \pm 1\big\}. \tag{8.19}$$

8.0.1. Analysis on $S^{\varepsilon}_{\omega,R,-}$.

In this section, we shall prove that any radial solution $\psi \in S^{\varepsilon}_{\omega,R,-}$ blows up forward in time.

PROPOSITION 8.0.5. *Assume that $0 < \varepsilon < \min\{\varepsilon_0(R_*), \varepsilon_*\}$, where $\varepsilon_0(R_*)$ is a constant given by Lemma 6.0.3. Let $R \in (\sqrt{2\varepsilon}, R_*)$, and let $\psi \in S^{\varepsilon}_{\omega,R,-}$. Then, the maximal lifespan $T_{\max}(\psi)$ is finite: the finite time blowup forward in time.*

PROOF OF PROPOSITION 8.0.5. We see from Lemma 8.0.3, Lemma 6.0.3 and Lemma 8.0.1 that there exist a time $T_0 \in [0, T_{\max}(\psi))$ and $\kappa_1(R_*) > 0$ such that

$$\inf_{t \in [T_0, T_{\max}(\psi))} |\mathcal{K}(\psi(t))| \ge \kappa_1(R_*). \tag{8.20}$$

Moreover, it follows from Lemma 8.0.2 that there exists a time $\tau \in [T_0, T_{\max}(\psi))$ such that

$$\widetilde{d}_\omega(\psi(\tau)) \geq \delta_X \geq \delta_*. \tag{8.21}$$

Hence, Lemma 8.0.4 together with $\mathfrak{S}(\psi) = -1$ shows that

$$\mathcal{K}(\psi(\tau)) < 0. \tag{8.22}$$

Putting (8.20) and (8.22) together, we find that

$$\sup_{t \in [T_0, T_{\max}(\psi))} \mathcal{K}(\psi(t)) \leq -\kappa_1(R_*). \tag{8.23}$$

Then, the same argument as the proof of Theorem 1.3 in [**2**] is available. Thus, we find that $T_{\max}(\psi) < \infty$. □

8.0.2. Analysis on $S^{\varepsilon}_{\omega,R,+}$. In this section, we shall prove that any solution $\psi \in S^{\varepsilon}_{\omega,R,+}$ scatters forward in time. Lemma 8.0.3 together with time-translation allows us to restrict ourselves to the solutions in $S^{\varepsilon}_{\omega,R_*,+}$. Then, Lemma 6.0.3 together with Lemma 8.0.1 determines constants $\varepsilon_0(R_*) > 0$ and $\kappa_1(R_*) > 0$ such that if $\varepsilon \in (0, \varepsilon_0(R_*))$, then any solution $\psi \in S^{\varepsilon}_{\omega,R_*,+}$ obeys

$$\inf_{t \in [0, T_{\max}(\psi))} |\mathcal{K}(\psi(t))| \geq \kappa_1(R_*). \tag{8.24}$$

Furthermore, this together with Lemma 8.0.2 and Lemma 8.0.4 shows that for any $\varepsilon \in (0, \varepsilon_0(R_*))$ and any $\psi \in S^{\varepsilon}_{\omega,R_*,+}$,

$$\inf_{[0, T_{\max}(\psi))} \mathcal{K}(\psi(t)) \geq \kappa_1(R_*). \tag{8.25}$$

It is well known (see, e.g, the claim (v) of Theorem 4.1 in [**3**]) that the boundedness of a solution ψ in the Strichartz-type space $W_{p+1}([0, T_{\max}(\psi)) \cap W_{2^*}([0, T_{\max}(\psi))$ implies the scattering forward in time. Thus, our aim is to prove the following:

PROPOSITION 8.0.6. *For any $\omega \in (0, \omega_*)$, there exists $\varepsilon \in (0, \varepsilon_*)$ such that for any $\psi \in S^{\varepsilon}_{\omega,R_*,+}$,*

$$\|\psi\|_{W_{p+1}([0,T_{\max}(\psi)) \cap W^{2^*}([0,T_{\max}(\psi)))} < \infty. \tag{8.26}$$

In particular, the solution ψ scatters forward in time.

PROOF OF PROPOSITION 8.0.6. We divide the proof into several parts: In section 8.0.2.1, we suppose for contradiction that the claim was false, and extract some sequence of non-scattering solutions in $S^{\varepsilon_*}_{\omega,R_*,+}$. In Section 8.0.2.2, we apply a profile decomposition to this sequence, and obtain "linear profiles". In Section 8.0.2.3, we introduce the "nonlinear profiles", and investigate their fundamental properties. Furthermore, in Section 8.0.2.4, we show the existence of a nonlinear profile whose Strichartz norm diverges. Finally, in Section 8.0.2.5, we derive a contradiction by showing the existence of the "critical element", and complete the proof.

8.0.2.1. *Setup.* For any $E > 0$, we define $\nu(E)$ by

(8.27)
$$\nu(E) := \sup \left\{ \|\psi\|_{W_{p+1}([0,T_{\max}(\psi)))\cap W_{2^*}([0,T_{\max}(\psi)))} : \psi \in S_{\omega,R_*,+}^{\varepsilon_*}, \ \mathcal{H}(\psi) \le E \right\}.$$

Furthermore, we put

(8.28)
$$E_* := \sup\{E > 0 \colon \nu(E) < \infty\}.$$

If $\psi \in S_{\omega,R_*,+}^{\varepsilon_*}$ and $\mathcal{H}(\psi) < \mathcal{H}(\Phi_\omega)$, then we see from (8.25) that $\psi \in PW_{\omega,+}$. Hence, it follows from Theorem 1.0.3 that

(8.29)
$$E_* \ge \mathcal{H}(\Phi_\omega).$$

Thus, what we want to prove is that $E_* > \mathcal{H}(\Phi_\omega)$. We prove this by contradiction, and therefore suppose that

(8.30)
$$E_* = \mathcal{H}(\Phi_\omega).$$

Then, we can take a sequence $\{\varepsilon_n\}$ of constants in $(0, \varepsilon_*)$ and a sequence $\{\psi_n\}$ of solutions such that

(8.31)
$$\lim_{n\to\infty} \varepsilon_n = 0,$$

(8.32)
$$\varepsilon_n < \min\left\{\varepsilon_0(R_*), \ \frac{\kappa_1(R_*)}{10d(p-1)+10}\right\},$$

(8.33)
$$\psi_n \in S_{\omega,R_*,+}^{\varepsilon_n},$$

(8.34)
$$\|\psi_n\|_{W_{p+1}([0,T_{\max}(\psi_n)))\cap W_{2^*}([0,T_{\max}(\psi_n)))} = \infty.$$

In particular, we have that

(8.35)
$$\inf_{n\ge1} \inf_{t\in[0,T_{\max}(\psi_n))} \widetilde{d}_\omega(\psi_n(t)) \ge R_*,$$

(8.36)
$$\lim_{n\to\infty} \mathcal{H}(\psi_n) = \mathcal{H}(\Phi_\omega) = E_*.$$

Furthermore, it follows from (8.25) and Lemma 8.0.4 that

(8.37)
$$\inf_{n\ge1} \inf_{t\in[0,T_{\max}(\psi_n))} \mathcal{K}(\psi_n(t)) \ge \kappa_1(R_*),$$

(8.38)
$$\sup_{n\ge1} \sup_{t\in[0,T_{\max}(\psi_n))} \|\psi_n(t)\|_{H^1} < \infty.$$

8.0.2.2. *Linear profiles.* Let $\{\psi_n\}$ be the sequence of solutions obtained in Section 8.0.2.1. We employ the profile decomposition for a sequence in the homogeneous space $\dot{H}^1(\mathbb{R}^d)$ (see Theorem 1.6 in [**19**] and Lemma 2.10 in [**21**]). We apply it to the sequence $\{|\nabla|^{-1}\langle\nabla\rangle\psi_n(0)\}$ in $\dot{H}^1(\mathbb{R}^d)$, instead of $\{\psi_n\}$ itself. Then, we find[1] that there exists a subsequence of $\{\psi_n\}$ (denoted by the same symbol $\{\psi_n\}$) with the following properties: there exist a family $\{\{(x_n^1, t_n^1, \lambda_n^1)\}, \{(x_n^2, t_n^2, \lambda_n^2)\}, \dots\}$ of sequences in $\mathbb{R}^d \times \mathbb{R} \times (0,\infty)$, a family $\{\widetilde{u}^1, \widetilde{u}^2, \dots\}$ of functions in $H^1(\mathbb{R}^d)$, and a family $\{\{w_n^1\}, \{w_n^2\}, \dots\}$ of sequences in $H^1(\mathbb{R}^d)$ such that:

[1]We do not use the assumption that every ψ_n is radial in this subsection (Section 8.0.2.2) and therefore translation parameters appear in the linear profile decomposition. In Section 8.0.2.4 and 8.0.2.5, we need the radial symmetry of the solutions so that we can take every translation $x_n^j = 0$.

- For any $j \ge 1$,

$$\lim_{n\to\infty} t_n^j = t_\infty^j \in \mathbb{R} \cup \{\pm\infty\}, \tag{8.39}$$

$$\lim_{n\to\infty} \lambda_n^j = \lambda_\infty^j \in \{0,1,\infty\}, \qquad \begin{cases} \lambda_n^j \equiv 1 & \text{if} \quad \lambda_\infty^j = 1, \\ \lambda_n^j \le 1 & \text{if} \quad \lambda_\infty^j = 0. \end{cases} \tag{8.40}$$

- For any $j' \neq j$,

$$\lim_{n\to\infty} \Big\{ \frac{\lambda_n^{j'}}{\lambda_n^j} + \frac{\lambda_n^j}{\lambda_n^{j'}} + \frac{|x_n^j - x_n^{j'}|}{\lambda_n^j} + \frac{|t_n^j - t_n^{j'}|}{(\lambda_n^j)^2} \Big\} = \infty. \tag{8.41}$$

- For any $\dot{H}^1$-admissible pair[2] (q,r),

$$\lim_{j\to\infty}\lim_{n\to\infty} \big\||\nabla|^{-1}\langle\nabla\rangle e^{it\Delta} w_n^j\big\|_{L_t^r L_x^q(\mathbb{R})} = 0. \tag{8.42}$$

- For any $n \ge 1$ and $k \ge 1$,

$$\begin{aligned} e^{it\Delta}\psi_n(0) &= \sum_{j=1}^k \langle\nabla\rangle^{-1}|\nabla| G_n^j\big(e^{it\Delta}|\nabla|^{-1}\langle\nabla\rangle\widetilde{u}^j\big) + e^{it\Delta} w_n^k \\ &= \sum_{j=1}^k \frac{\langle\nabla\rangle^{-1}\langle\lambda_n^j\nabla\rangle}{\lambda_n^j} e^{i(t-t_n^j)\Delta} g_n^j \widetilde{u}^j + e^{it\Delta} w_n^k, \end{aligned} \tag{8.43}$$

where G_n^j and g_n^j are the operators defined by

$$G_n^j v(x,t) := \frac{1}{(\lambda_n^j)^{\frac{d-2}{2}}} v\Big(\frac{x-x_n^j}{\lambda_n^j}, \frac{t-t_n^j}{(\lambda_n^j)^2}\Big), \tag{8.44}$$

$$g_n^j u(x) := \frac{1}{(\lambda_n^j)^{\frac{d-2}{2}}} u\Big(\frac{x-x_n^j}{\lambda_n^j}\Big). \tag{8.45}$$

Note that

$$\frac{\langle\nabla\rangle^{-1}\langle\lambda_n^j\nabla\rangle}{\lambda_n^j} e^{i(t-t_n^j)\Delta} g_n^j\widetilde{u}^j = g_n^j\sigma_n^j e^{i\frac{t-t_n^j}{(\lambda_n^j)^2}\Delta}\widetilde{u}^j = G_n^j\sigma_n^j e^{it\Delta}\widetilde{u}^j, \tag{8.46}$$

where

$$\sigma_n^j := \frac{\langle(\lambda_n^j)^{-1}\nabla\rangle^{-1}\langle\nabla\rangle}{\lambda_n^j}. \tag{8.47}$$

- For any $k \ge 1$ and any $1 \le j \le k$,

$$\lim_{n\to\infty} \langle\nabla\rangle^{-1}|\nabla| e^{-i\frac{t_n^j}{(\lambda_n^j)}\Delta}(g_n^j)^{-1}|\nabla|^{-1}\langle\nabla\rangle w_n^k = 0 \quad \text{weakly in } H^1(\mathbb{R}^d). \tag{8.48}$$

- For any $k \ge 1$,

$$\lim_{n\to\infty} \Big\{ \|\langle\nabla\rangle\psi_n(0)\|_{L^2}^2 - \sum_{j=1}^k \|\langle\nabla\rangle g_n^j\sigma_n^j e^{-i\frac{t_n^j}{(\lambda_n^j)^2}\Delta}\widetilde{u}^j\|_{L^2}^2 - \|\langle\nabla\rangle w_n^k\|_{L^2}^2 \Big\} = 0. \tag{8.49}$$

---

[2] A pair (q,r) is said to be $\dot{H}^1$-admissible if $\frac{1}{r} = \frac{d}{2}(\frac{1}{2} - \frac{1}{q} - \frac{1}{d})$ and $(q,r) \in [2,\infty] \times [2,\infty]$.

Furthermore, the following hold for all $k \ge 1$ (see Lemma 2.2. in [**3**]):

(8.50) $$\lim_{n\to\infty} \Big\{\mathcal{M}(\psi_n) - \sum_{j=1}^{k} \mathcal{M}\big(g_n^j \sigma_n^j e^{-i\frac{t_n^j}{(\lambda_n^j)^2}\Delta}\widetilde{u}^j\big) - \mathcal{M}(w_n^k)\Big\} = 0,$$

(8.51) $$\lim_{n\to\infty} \Big\{\mathcal{H}(\psi_n) - \sum_{j=1}^{k} \mathcal{H}\big(g_n^j \sigma_n^j e^{-i\frac{t_n^j}{(\lambda_n^j)^2}\Delta}\widetilde{u}^j\big) - \mathcal{H}(w_n^k)\Big\} = 0,$$

(8.52) $$\lim_{n\to\infty} \Big\{\mathcal{S}_\omega(\psi_n) - \sum_{j=1}^{k} \mathcal{S}_\omega\big(g_n^j \sigma_n^j e^{-i\frac{t_n^j}{(\lambda_n^j)^2}\Delta}\widetilde{u}^j\big) - \mathcal{S}_\omega(w_n^k)\Big\} = 0,$$

(8.53) $$\lim_{n\to\infty} \Big\{\mathcal{I}_\omega(\psi_n(0)) - \sum_{j=1}^{k} \mathcal{I}_\omega\big(g_n^j \sigma_n^j e^{-i\frac{t_n^j}{(\lambda_n^j)^2}\Delta}\widetilde{u}^j\big) - \mathcal{I}_\omega(w_n^k)\Big\} = 0.$$

Note that it follows from Strichartz' estimate, (8.38) and (8.49) that for any $k \ge 1$, there exists a number $N(k)$ such that for any $n \ge N(k)$,

(8.54) $$\big\|\langle\nabla\rangle e^{it\Delta} w_n^k\big\|_{St(\mathbb{R})} \lesssim \|w_n^k\|_{H^1} \lesssim 1.$$

Moreover, it follows from (8.42) and (8.54) that for any $1 + \frac{4}{d} < q \le 2^* - 1$,

(8.55) $$\begin{aligned}\lim_{k\to\infty}\lim_{n\to\infty} \|e^{it\Delta} w_n^k\|_{W_{q+1}(\mathbb{R})} &\le \lim_{k\to\infty}\lim_{n\to\infty} \|e^{it\Delta} w_n^k\|_{W_{2+\frac{4}{d}}(\mathbb{R})}^{1-s_q} \|e^{it\Delta} w_n^k\|_{W_{2^*}(\mathbb{R})}^{s_q} \\ &\lesssim \lim_{k\to\infty}\lim_{n\to\infty} \||\nabla|^{-1}\langle\nabla\rangle e^{it\Delta} w_n^k\|_{W_{2^*}(\mathbb{R})}^{s_q} = 0.\end{aligned}$$

We shall discuss fundamental properties of operators g_n^j (see (8.45)) and σ_n^j (see (8.47)). To this end, for each $j \ge 1$, we introduce an operator σ_∞^j as

(8.56) $$\sigma_\infty^j := \begin{cases} |\nabla|^{-1}\langle\nabla\rangle & \text{if} \quad \lambda_\infty^j = 0, \\ 1 & \text{if} \quad \lambda_\infty^j = 1, \\ 0 & \text{if} \quad \lambda_\infty^j = \infty. \end{cases}$$

LEMMA 8.0.7. *The following hold for all function $f \in H^1(\mathbb{R}^d)$ and all $j \ge 1$:*

(i)

(8.57) $$\lim_{n\to\infty} \big\|\nabla\{\sigma_n^j f - \sigma_\infty^j f\}\big\|_{L^2} = 0;$$

(ii)

(8.58) $$\lim_{n\to\infty} \big\|g_n^j \sigma_n^j f\big\|_{L^2} = \lim_{n\to\infty} \big\|\lambda_n^j \sigma_n^j f\big\|_{L^2} = \begin{cases} 0 & \textit{if} \quad \lambda_\infty^j = 0, \\ \|f\|_{L^2} & \textit{if} \quad \lambda_\infty^j = 1, \\ \|\langle\nabla\rangle f\|_{L^2} & \textit{if} \quad \lambda_\infty^j = \infty. \end{cases}$$

PROOF OF LEMMA 8.0.7. We shall prove (8.57). If $\lambda_\infty^j = 1$, then $\sigma_n^j \equiv 1$ and therefore the claim is trivial. Assume that $\lambda_\infty^j = 0$. Then, we see from Parseval's identity that

$$\begin{aligned}\lim_{n\to\infty}\||\nabla|\{\sigma_n^j f-\sigma_\infty^j f\}\|_{L^2}&=\lim_{n\to\infty}\left\||\nabla|\left\{\frac{\langle(\lambda_n^j)^{-1}\nabla\rangle^{-1}}{\lambda_n^j}-|\nabla|^{-1}\right\}\langle\nabla\rangle f\right\|_{L^2}\\&=\lim_{n\to\infty}\left\|\left\{\frac{|\xi|}{\sqrt{(\lambda_n^j)^2+|\xi|^2}}-1\right\}\langle\xi\rangle\mathcal{F}[f]\right\|_{L^2_\xi}.\end{aligned}\tag{8.59}$$

Moreover, since $\lim_{n\to\infty}\lambda_n^j=\lambda_\infty^j=0$, we see that

$$\lim_{n\to\infty}\left\{\frac{|\xi|}{\sqrt{(\lambda_n^j)^2+|\xi|^2}}-1\right\}\langle\xi\rangle\mathcal{F}[f](\xi)=0 \qquad \text{for almost all } \xi\in\mathbb{R}^d.\tag{8.60}$$

Hence, we find from Lebesgue's convergence theorem that (8.57) holds. Similarly, we can prove the case where $\lambda_\infty^j=\infty$.

Next, we give a proof of (8.58). The case where $\lambda_\infty^j=1$ is trivial. Assume that $\lambda_\infty^j=0$. Then, we see from the substitution of variables and Parseval's identity that

$$\lim_{n\to\infty}\|g_n^j\sigma_n^j f\|_{L^2}=\lim_{n\to\infty}\|\lambda_n^j\sigma_n^j f\|_{L^2}=\lim_{n\to\infty}\left\|\frac{1}{\sqrt{1+\frac{|\xi|^2}{(\lambda_n^j)^2}}}\langle\xi\rangle\mathcal{F}[f]\right\|_{L^2_\xi}.\tag{8.61}$$

Since $\lim_{n\to\infty}\lambda_n^j=\lambda_\infty^j=0$, we find from (8.61) and Lebesgue's convergence theorem that (8.58) holds. Similarly, we can prove the case where $\lambda_\infty^j=\infty$. □

LEMMA 8.0.8. *Let j be a number such that $\lambda_\infty^j\neq\infty$. Then, for any $n\ge1$, any $q\in[2,\infty)$ and any function f with $\langle\nabla\rangle f\in L^q(\mathbb{R}^d)$,*

$$\|\lambda_n^j\sigma_n^j f\|_{L^q}+\|\nabla\sigma_n^j f\|_{L^q}\lesssim\|\langle\nabla\rangle f\|_{L^q},\tag{8.62}$$

where the implicit constant depends only on d and q.

PROOF OF LEMMA 8.0.8. If $\lambda_\infty^j=1$, then the claim is trivial. When $\lambda_\infty^j=0$, we find from the Mihlin multiplier theorem that

$$\begin{aligned}\|\lambda_n^j\sigma_n^j f\|_{L^q}+\|\nabla\sigma_n^j f\|_{L^q}&\lesssim\|\langle\nabla\rangle f\|_{L^q}+\|\sigma_n^j(\sigma_\infty^j)^{-1}|\nabla|\sigma_\infty^j f\|_{L^q}\\&\lesssim\|\langle\nabla\rangle f\|_{L^q}+\||\nabla|\sigma_\infty^j f\|_{L^q}\lesssim\|\langle\nabla\rangle f\|_{L^q}.\end{aligned}\tag{8.63}$$

Thus, we have proved the lemma. □

LEMMA 8.0.9. *Let $1+\frac{4}{d}<q<2^*-1$, and let $j\ge1$ be a number for which $\lambda_\infty^j\in\{0,\infty\}$. Then, the linear profile $\tilde{u}^j$ satisfies*

$$\lim_{n\to\infty}\|G_n^j\sigma_n^j e^{it\Delta}\widetilde{u}^j|_{t=0}\|_{L^{q+1}}=\lim_{n\to\infty}\|g_n^j\sigma_n^j e^{-i\frac{t_n^j}{(\lambda_n^j)^2}\Delta}\widetilde{u}^j\|_{L^{q+1}}=0.\tag{8.64}$$

PROOF OF LEMMA 8.0.9. Assume that $\lambda_\infty^j\in\{0,\infty\}$. Then, we see from the Gagliardo-Nirenberg inequality that

$$\|g_n^j\sigma_n^j e^{-i\frac{t_n^j}{(\lambda_n^j)^2}\Delta}\widetilde{u}^j\|_{L^{q+1}}^{q+1}\lesssim\|\lambda_n^j\sigma_n^j\widetilde{u}^j\|_{L^2}^{q+1-\frac{d(q-1)}{2}}\|\nabla\sigma_n^j\widetilde{u}^j\|_{L^2}^{\frac{d(q-1)}{2}}.\tag{8.65}$$

If $\lambda_\infty^j=0$, then it follows from (8.65) and Lemma 8.0.7 that

$$\lim_{n\to\infty}\|g_n^j\sigma_n^j e^{-i\frac{t_n^j}{(\lambda_n^j)^2}\Delta}\widetilde{u}^j\|_{L^{q+1}}^{q+1}\lesssim\lim_{n\to\infty}\|\lambda_n^j\sigma_n^j\widetilde{u}^j\|_{L^2}^{q+1-\frac{d(q-1)}{2}}\|\langle\nabla\rangle\widetilde{u}^j\|_{L^2}^{\frac{d(q-1)}{2}}=0.\tag{8.66}$$

On the other hand, if $\lambda_\infty^j = \infty$, then it follows from (8.65) and Lemma 8.0.7 that
(8.67)

$$\lim_{n\to\infty} \left\| g_n^j \sigma_n^j e^{-i\frac{t_n^j}{(\lambda_n^j)^2}\Delta} \widetilde{u}^j \right\|_{L^{q+1}}^{q+1} \lesssim \left\| \langle \nabla \rangle \widetilde{u}^j \right\|_{L^2}^{q+1-\frac{d(q-1)}{2}} \lim_{n\to\infty} \left\| \nabla \sigma_n^j \widetilde{u}^j \right\|_{L^2}^{\frac{d(q-1)}{2}} = 0.$$

Thus, we have completed the proof. □

8.0.2.3. *Nonlinear profiles.* In this subsection, we introduce the "nonlinear profile" associated with $(\widetilde{u}^j, \{(x_n^j, t_n^j, \lambda_n^j)\})$ for each $j \geq 1$. From the point of view of (8.43), we first introduce a function U_n^j as the solution to (NLS) satisfying

$$U_n^j(0) = \frac{\langle \nabla \rangle^{-1} \langle \lambda_n^j \nabla \rangle}{\lambda_n^j} e^{-it_n^j \Delta} g_n^j \widetilde{u}^j. \tag{8.68}$$

Then, we find from (8.46) that

$$e^{it\Delta} U_n^j(0) = G_n^j \sigma_n^j e^{it\Delta} \widetilde{u}^j. \tag{8.69}$$

Furthermore, Duhamel's formula for U_n^j together with (8.69) shows that U_n^j satisfies

$$U_n^j(t) - G_n^j \sigma_n^j e^{it\Delta} \widetilde{u}^j = i \int_0^t e^{i(t-s)\Delta} \left\{ F^\dagger[U_n^j](s) + F^\ddagger[U_n^j](s) \right\} ds, \tag{8.70}$$

where see (1.65) for the definitions of $F^\dagger$ and $F^\ddagger$. Applying $(\sigma_n^j)^{-1}(G_n^j)^{-1}$ to the both sides in (8.70), we find that

(8.71)

$$\begin{aligned}
&(\sigma_n^j)^{-1}(G_n^j)^{-1} U_n^j(t) - e^{it\Delta} \widetilde{u}^j \\
&= i(\sigma_n^j)^{-1}(g_n^j)^{-1} \int_{-\frac{t_n^j}{(\lambda_n^j)^2}}^t e^{i(\lambda_n^j)^2(t-s)\Delta} \left\{ (\lambda_n^j)^{-\frac{(d-2)(p-1)}{2}} g_n^j F^\dagger[(G_n^j)^{-1} U_n^j](s) \right. \\
&\qquad \left. + (\lambda_n^j)^{-2} g_n^j F^\ddagger[(G_n^j)^{-1} U_n^j](s) \right\} (\lambda_n^j)^2 \, ds \\
&= i(\sigma_n^j)^{-1} \int_{-\frac{t_n^j}{(\lambda_n^j)^2}}^t e^{i(t-s)\Delta} \left\{ (\lambda_n^j)^{\frac{4-(d-2)(p-1)}{2}} F^\dagger[\sigma_n^j (\sigma_n^j)^{-1} (G_n^j)^{-1} U_n^j](s) \right. \\
&\qquad \left. + F^\ddagger[\sigma_n^j (\sigma_n^j)^{-1} (G_n^j)^{-1} U_n^j](s) \right\} ds.
\end{aligned}$$

From the point of view of (8.71), for each number $j \geq 1$ with $\lambda_\infty^j \in \{0, 1\}$, we define the nonlinear profile $\widetilde{\psi}^j$ associated with $(\widetilde{u}^j, \{x_n^j, t_n^j, \lambda_n^j\})$ as the solution to

$$\begin{aligned}
\widetilde{\psi}^j(t) = e^{it\Delta} \widetilde{u}^j + i \int_{-\frac{t_\infty^j}{(\lambda_\infty^j)^2}}^t e^{i(t-s)\Delta} (\sigma_\infty^j)^{-1} \Big\{ & (\lambda_\infty^j)^{\frac{4-(d-2)(p-1)}{2}} F^\dagger[\sigma_\infty^j \widetilde{\psi}^j](s) \\
& + F^\ddagger[\sigma_\infty^j \widetilde{\psi}^j](s) \Big\} ds.
\end{aligned} \tag{8.72}$$

When $-\frac{t_\infty^j}{(\lambda_\infty^j)^2} \in \{\pm\infty\}$, the equation (8.72) is interpreted as the final value problem at $\pm\infty$. Moreover, if $j \geq 1$ is a number for which $\lambda_\infty^j = \infty$, then we define $\widetilde{\psi}^j(t) := e^{it\Delta} \widetilde{u}^j$ (we define $\widetilde{\psi}^j$ as a free solution to (NLS) so that (8.76) holds; in particular, if $\widetilde{u}^j \equiv 0$, then $\widetilde{\psi}^j \equiv 0$).

We simply write $I^j_{\max} := I_{\max}(\widetilde{\psi}^j)$, $T^j_{\max} := T_{\max}(\widetilde{\psi}^j)$ and $T^j_{\min} := T_{\min}(\widetilde{\psi}^j)$. Moreover, for an interval I and a number $j \ge 1$, we define

$$\mathcal{W}^j(I) := \begin{cases} W_{p+1}(I) \cap W_{2^*}(I) & \text{if} \quad \lambda^j_\infty \in \{1,\infty\}, \\ W_{2^*}(I) & \text{if} \quad \lambda^j_\infty = 0. \end{cases} \tag{8.73}$$

We find from the construction of the nonlinear profiles (see (8.72)) that the following hold for all numbers $j \ge 1$:

- $\widetilde{\psi}^j \in C(I^j_{\max}, H^1(\mathbb{R}^d))$.
- $\widetilde{\psi}^j$ satisfies

$$\begin{cases} i\dfrac{\partial \sigma^j_\infty \widetilde{\psi}^j}{\partial t} + \Delta \sigma^j_\infty \widetilde{\psi}^j + F[\sigma^j_\infty \widetilde{\psi}^j] = 0 \quad \text{if} \quad \lambda^j_\infty = 1, \\ i\dfrac{\partial \sigma^j_\infty \widetilde{\psi}^j}{\partial t} + \Delta \sigma^j_\infty \widetilde{\psi}^j + F^\ddagger[\sigma^j_\infty \widetilde{\psi}^j] = 0 \quad \text{if} \quad \lambda^j_\infty = 0, \\ i\dfrac{\partial \widetilde{\psi}^j}{\partial t} + \Delta \widetilde{\psi}^j = 0 \quad \text{if} \quad \lambda^j_\infty = \infty. \end{cases} \tag{8.74}$$

- There exists a number $N(j)$ such that for any $n \ge N(j)$,

$$-\frac{t^j_n}{(\lambda^j_n)^2} \in I^j_{\max} \tag{8.75}$$

and

$$\lim_{n\to\infty} \Big\| \widetilde{\psi}^j\Big(-\frac{t^j_n}{(\lambda^j_n)^2} \Big) - e^{-i\frac{t^j_n}{(\lambda^j_n)^2}\Delta} \widetilde{u}^j \Big\|_{H^1} = 0. \tag{8.76}$$

- If $-\frac{t^j_\infty}{(\lambda^j_\infty)^2} = \infty$, then $T^j_{\max} = \infty$ and for any $T > T^j_{\min}$,

$$\| \sigma^j_\infty \widetilde{\psi}^j \|_{\mathcal{W}^j([T,\infty))} < \infty; \tag{8.77}$$

- If $-\frac{t^j_\infty}{(\lambda^j_\infty)^2} = -\infty$, then $T^j_{\min} = -\infty$ and for any $T < T^j_{\max}$,

$$\| \sigma^j_\infty \widetilde{\psi}^j \|_{\mathcal{W}^j((-\infty,T])} < \infty. \tag{8.78}$$

When $\lambda^j_\infty \in \{0,\infty\}$, it is convenient rewriting (8.76) in the form

$$\begin{cases} \lim\limits_{n\to\infty} \Big\| \widetilde{\psi}^j\Big(\cdot - x^j_n, -\dfrac{t^j_n}{(\lambda^j_n)^2} \Big) - g^j_n \sigma^j_n e^{-i\frac{t^j_n}{(\lambda^j_n)^2}\Delta} \widetilde{u}^j \Big\|_{H^1} = 0 \quad \text{if} \quad \lambda^j_\infty = 1, \\ \lim\limits_{n\to\infty} \Big\| \nabla g^j_n \Big\{ \sigma^j_\infty \widetilde{\psi}^j\Big(-\dfrac{t^j_n}{(\lambda^j_n)^2} \Big) - \sigma^j_n e^{-i\frac{t^j_n}{(\lambda^j_n)^2}\Delta} \widetilde{u}^j \Big\} \Big\|_{L^2} = 0 \quad \text{if} \quad \lambda^j_\infty = 0. \end{cases} \tag{8.79}$$

Now, for each number $j \ge 1$, we define

$$\psi^j_n := G^j_n \sigma^j_n \widetilde{\psi}^j, \tag{8.80}$$

and let $I_{\max,n}^{j}$ be the maximal existence-interval of ψ_n^j. Then, we see from (8.74) that

$$\left\{\begin{array}{ll} i\dfrac{\partial \psi_n^j}{\partial t}+\Delta\psi_n^j+F[\psi_n^j]=0 & \text{if} \quad \lambda_\infty^j=1,\\ i\dfrac{\partial \psi_n^j}{\partial t}+\Delta\psi_n^j+(\sigma_\infty^j)^{-1}F^{\ddagger}[G_n^j\sigma_\infty^j\widetilde{\psi}^j]=0 & \text{if} \quad \lambda_\infty^j=0,\\ i\dfrac{\partial \psi_n^j}{\partial t}+\Delta\psi_n^j=0 & \text{if} \quad \lambda_\infty^j=\infty, \end{array}\right. \tag{8.81}$$

and

$$I_{\max,n}^{j}=\left\{\begin{array}{ll} ((\lambda_n^j)^2T_{\min}^j+t_n^j,\ (\lambda_n^j)^2T_{\max}^j+t_n^j) & \text{if} \quad \lambda_\infty^j\in\{0,1\},\\ (-\infty,\infty) & \text{if} \quad \lambda_\infty^j=\infty. \end{array}\right. \tag{8.82}$$

Furthermore, (8.75) shows that $0\in I_{\max,n}^{j}$ for any $n\ge N(j)$.

In what follows, we discuss fundamental properties of nonlinear profiles in a series of lemmas. In particular, we show that there is at most one "bad nonlinear profile"; the other profiles are "good" in the sense of (8.135) and (8.136).

LEMMA 8.0.10. *Let j be a number for which $\lambda_\infty^j=0$, and let I be an interval on which*

$$\|\langle\nabla\rangle\widetilde{\psi}^j\|_{St(I)}<\infty. \tag{8.83}$$

Then, the following hold:

$$\lim_{n\to\infty}\|\sigma_\infty^j\widetilde{\psi}^j-\sigma_n^j\widetilde{\psi}^j\|_{W_{2^*}(I)}\lesssim\lim_{n\to\infty}\|\nabla\{\sigma_\infty^j\widetilde{\psi}^j-\sigma_n^j\widetilde{\psi}^j\}\|_{V_{2^*}(I)}=0, \tag{8.84}$$

$$\lim_{n\to\infty}\|\nabla\{\sigma_\infty^j\widetilde{\psi}^j-\sigma_n^j\widetilde{\psi}^j\}\|_{V_{2+\frac{4}{d}}(I)}=0, \tag{8.85}$$

$$\lim_{n\to\infty}\|\lambda_n^j\sigma_n^j\widetilde{\psi}^j\|_{V_{2+\frac{4}{d}}(I)}=0. \tag{8.86}$$

PROOF OF LEMMA 8.0.10. The inequality in (8.84) follows from Sobolev's embedding. We see from Hölder's inequality, Lemma 8.0.7, Lemma 8.0.8 and the assumption (8.83) that for almost all $t\in I$,

$$\begin{split} &\lim_{n\to\infty}\|\nabla\{\sigma_\infty^j\widetilde{\psi}^j(t)-\sigma_n^j\widetilde{\psi}^j(t)\}\|_{L^{\frac{2d(d+2)}{d^2+4}}}^{\frac{2(d+2)}{d-2}}\\ &\le\lim_{n\to\infty}\|\nabla\{\sigma_\infty^j\widetilde{\psi}^j(t)-\sigma_n^j\widetilde{\psi}^j(t)\}\|_{L^2(I)}^{\frac{8}{d-2}}\|\langle\nabla\rangle\widetilde{\psi}^j(t)\|_{L^{2^*}(I)}^2=0. \end{split} \tag{8.87}$$

Furthermore, we see from Lemma 8.0.8 and the assumption (8.83) that

$$\|\nabla\{\sigma_\infty^j\widetilde{\psi}^j(t)-\sigma_n^j\widetilde{\psi}^j(t)\}\|_{L^{\frac{2d(d+2)}{d^2+4}}}^{\frac{2(d+2)}{d-2}}\lesssim\|\langle\nabla\rangle\widetilde{\psi}^j(t)\|_{L^{\frac{2d(d+2)}{d^2+4}}}^{\frac{2(d+2)}{d-2}}\in L^1(I). \tag{8.88}$$

Thus, Lebesgue's convergence theorem shows that the equality in (8.84) holds. Similarly, we can prove (8.85). It remains to prove (8.86). We see from Hölder's inequality, Lemma 8.0.7, Lemma 8.0.8 and the assumption (8.83) that for almost all $t\in I$,

$$\lim_{n\to\infty}\|\lambda_n^j\sigma_n^j\widetilde{\psi}^j(t)\|_{L^{\frac{2(d+2)}{d}}}^{\frac{2(d+2)}{d}}\le\lim_{n\to\infty}\|\lambda_n^j\sigma_n^j\widetilde{\psi}^j(t)\|_{L^2(I)}^{\frac{4}{d}}\|\langle\nabla\rangle\widetilde{\psi}^j(t)\|_{L^{2^*}(I)}^2=0. \tag{8.89}$$

Furthermore, we see from Lemma 8.0.8 and the assumption (8.83) that

$$\|\lambda_n^j \sigma_n^j \widetilde{\psi}^j(t)\|_{L^{\frac{2(d+2)}{d}}}^{\frac{2(d+2)}{d}} \lesssim \|\langle\nabla\rangle \widetilde{\psi}^j(t)\|_{L^{\frac{2(d+2)}{d}}}^{\frac{2(d+2)}{d}} \in L^1(I). \tag{8.90}$$

Thus, Lebesgue's convergence theorem shows that (8.86) holds. □

LEMMA 8.0.11. *Let j be a number for which $\lambda_\infty^j = 0$, and let I be an interval on which*

$$\big\|\langle\nabla\rangle \widetilde{\psi}^j\big\|_{St(I)} < \infty. \tag{8.91}$$

Then, the nonlinear term in (8.81) *satisfies*

$$\lim_{n\to\infty} \big\|\langle\nabla\rangle\{(\sigma_\infty^j)^{-1} F^\ddagger[G_n^j \sigma_\infty^j \widetilde{\psi}^j] - F[\psi_n^j]\}\big\|_{L_{t,x}^{\frac{2(d+2)}{d+4}}(I_n^j)} = 0, \tag{8.92}$$

where $I_n^j := ((\lambda_n^j)^2 \inf I + t_n^j,\ (\lambda_n^j)^2 \sup I + t_n^j)$.

PROOF OF LEMMA 8.0.11. Note first that

$$F[\psi_n^j] = F^\ddagger[G_n^j \sigma_n^j \widetilde{\psi}^j] + F^\dagger[G_n^j \sigma_n^j \widetilde{\psi}^j]. \tag{8.93}$$

Next, observe that

(8.94)
$$\begin{aligned}
&\lim_{n\to\infty} \big\|\langle\nabla\rangle\{(\sigma_\infty^j)^{-1} F^\ddagger[G_n^j \sigma_\infty^j \widetilde{\psi}^j] - F^\ddagger[G_n^j \sigma_n^j \widetilde{\psi}^j]\}\big\|_{L_{t,x}^{\frac{2(d+2)}{d+4}}(I_n^j)} \\
&= \lim_{n\to\infty} (\lambda_n^j)^{-2} \big\|G_n^j (\lambda_n^j)^{-1} |\nabla| F^\ddagger[\sigma_\infty^j \widetilde{\psi}^j] - G_n^j \langle(\lambda_n^j)^{-1}\nabla\rangle F^\ddagger[\sigma_n^j \widetilde{\psi}^j]\big\|_{L_{t,x}^{\frac{2(d+2)}{d+4}}(I_n^j)} \\
&= \lim_{n\to\infty} \big\||\nabla| F^\ddagger[\sigma_\infty^j \widetilde{\psi}^j] - \lambda_n^j \langle(\lambda_n^j)^{-1}\nabla\rangle F^\ddagger[\sigma_n^j \widetilde{\psi}^j]\big\|_{L_{t,x}^{\frac{2(d+2)}{d+4}}(I)} \\
&\lesssim \lim_{n\to\infty} \big\|\nabla\{F^\ddagger[\sigma_\infty^j \widetilde{\psi}^j] - F^\ddagger[\sigma_n^j \widetilde{\psi}^j]\}\big\|_{L_{t,x}^{\frac{2(d+2)}{d+4}}(I)} \\
&\quad + \lim_{n\to\infty} \lambda_n^j \big\|\{|(\lambda_n^j)^{-1}\nabla| - \langle(\lambda_n^j)^{-1}\nabla\rangle\} F^\ddagger[\sigma_n^j \widetilde{\psi}^j]\big\|_{L_{t,x}^{\frac{2(d+2)}{d+4}}(I)}.
\end{aligned}$$

We consider the first term on the right-hand side of (8.94). We see from elementary computations and Hölder's inequality that

(8.95)
$$\begin{aligned}
&\lim_{n\to\infty} \big\|\nabla\{F^\ddagger[\sigma_\infty^j \widetilde{\psi}^j] - F^\ddagger[\sigma_n^j \widetilde{\psi}^j]\}\big\|_{L_{t,x}^{\frac{2(d+2)}{d+4}}(I)} \\
&\le \lim_{n\to\infty} \big\||\sigma_\infty^j \widetilde{\psi}^j|^{\frac{4}{d-2}} \nabla \sigma_\infty^j \widetilde{\psi}^j - |\sigma_n^j \widetilde{\psi}^j|^{\frac{4}{d-2}} \nabla \sigma_n^j \widetilde{\psi}^j\}\big\|_{L_{t,x}^{\frac{2(d+2)}{d+4}}(I)} \\
&\quad + \lim_{n\to\infty} \big\||\sigma_\infty^j \widetilde{\psi}^j|^{\frac{8-2d}{d-2}} (\sigma_\infty^j \widetilde{\psi}^j)^2 \overline{\nabla \sigma_\infty^j \widetilde{\psi}^j} - |\sigma_n^j \widetilde{\psi}^j|^{\frac{8-2d}{d-2}} (\sigma_n^j \widetilde{\psi}^j)^2 \overline{\nabla \sigma_n^j \widetilde{\psi}^j}\}\big\|_{L_{t,x}^{\frac{2(d+2)}{d+4}}(I)}.
\end{aligned}$$

Furthermore, we see from the triangle inequality, Hölder's inequality, Sobolev's embedding, Lemma 8.0.8, Lemma 8.0.10 and the assumption (8.91) that if $d \ge 6$

(hence $\frac{4}{d-2} \le 1$), then

(8.96)
$$
\begin{aligned}
&\lim_{n\to\infty} \big\| |\sigma^j_\infty \widetilde{\psi}^j|^{\frac{4}{d-2}} \nabla \sigma^j_\infty \widetilde{\psi}^j - |\sigma^j_n \widetilde{\psi}^j|^{\frac{4}{d-2}} \nabla \sigma^j_n \widetilde{\psi}^j \}\big\|_{L^{\frac{2(d+2)}{d+4}}_{t,x}(I)} \\
&\quad + \lim_{n\to\infty} \big\| |\sigma^j_\infty \widetilde{\psi}^j|^{\frac{8-2d}{d-2}} (\sigma^j_\infty \widetilde{\psi}^j)^2 \overline{\nabla \sigma^j_\infty \widetilde{\psi}^j} - |\sigma^j_n \widetilde{\psi}^j|^{\frac{8-2d}{d-2}} (\sigma^j_n \widetilde{\psi}^j)^2 \overline{\nabla \sigma^j_n \widetilde{\psi}^j} \}\big\|_{L^{\frac{2(d+2)}{d+4}}_{t,x}(I)} \\
&\lesssim \lim_{n\to\infty} \big\| |\sigma^j_\infty \widetilde{\psi}^j|^{\frac{4}{d-2}} \nabla \sigma^j_\infty \widetilde{\psi}^j - |\sigma^j_n \widetilde{\psi}^j|^{\frac{4}{d-2}} \nabla \sigma^j_\infty \widetilde{\psi}^j \}\big\|_{L^{\frac{2(d+2)}{d+4}}_{t,x}(I)} \\
&\quad + \lim_{n\to\infty} \big\| |\sigma^j_\infty \widetilde{\psi}^j|^{\frac{8-2d}{d-2}} (\sigma^j_\infty \widetilde{\psi}^j)^2 \overline{\nabla \sigma^j_\infty \widetilde{\psi}^j} - |\sigma^j_n \widetilde{\psi}^j|^{\frac{8-2d}{d-2}} (\sigma^j_n \widetilde{\psi}^j)^2 \overline{\nabla \sigma^j_\infty \widetilde{\psi}^j} \}\big\|_{L^{\frac{2(d+2)}{d+4}}_{t,x}(I)} \\
&\quad + \lim_{n\to\infty} \big\| |\sigma^j_n \widetilde{\psi}^j|^{\frac{4}{d-2}} \nabla \sigma^j_\infty \widetilde{\psi}^j - |\sigma^j_n \widetilde{\psi}^j|^{\frac{4}{d-2}} \nabla \sigma^j_n \widetilde{\psi}^j \}\big\|_{L^{\frac{2(d+2)}{d+4}}_{t,x}(I)} \\
&\quad + \lim_{n\to\infty} \big\| |\sigma^j_n \widetilde{\psi}^j|^{\frac{8-2d}{d-2}} (\sigma^j_n \widetilde{\psi}^j)^2 \overline{\nabla \sigma^j_\infty \widetilde{\psi}^j} - |\sigma^j_n \widetilde{\psi}^j|^{\frac{8-2d}{d-2}} (\sigma^j_n \widetilde{\psi}^j)^2 \overline{\nabla \sigma^j_n \widetilde{\psi}^j} \}\big\|_{L^{\frac{2(d+2)}{d+4}}_{t,x}(I)} \\
&\lesssim \lim_{n\to\infty} \|\sigma^j_\infty \widetilde{\psi}^j - \sigma^j_n \widetilde{\psi}^j\|^{\frac{4}{d-2}}_{W_{2^*}(I)} \|\nabla \sigma^j_\infty \widetilde{\psi}^j\|_{V_{2+\frac{4}{d}}(I)} \\
&\quad + \lim_{n\to\infty} \|\sigma^j_n \widetilde{\psi}^j\|^{\frac{4}{d-2}}_{W_{2^*}(I)} \|\nabla \{\sigma^j_\infty \widetilde{\psi}^j - \sigma^j_\infty \widetilde{\psi}^j\}\|_{V_{2+\frac{4}{d}}(I)} \\
&\lesssim \lim_{n\to\infty} \|\sigma^j_\infty \widetilde{\psi}^j - \sigma^j_n \widetilde{\psi}^j\|^{\frac{4}{d-2}}_{W_{2^*}(I)} \|\langle\nabla\rangle \widetilde{\psi}^j\|_{V_{2+\frac{4}{d}}(I)} \\
&\quad + \lim_{n\to\infty} \|\langle\nabla\rangle \widetilde{\psi}^j\|^{\frac{4}{d-2}}_{V_{2^*}(I)} \|\nabla \{\sigma^j_\infty \widetilde{\psi}^j - \sigma^j_\infty \widetilde{\psi}^j\}\|_{V_{2+\frac{4}{d}}(I)} \\
&= 0.
\end{aligned}
$$

Similarly, we can verify that if $3 \le d \le 5$ (hence $1 < \frac{4}{d-2} = 1 + \frac{6-d}{d-2}$), then

(8.97)
$$
\begin{aligned}
&\lim_{n\to\infty} \big\| |\sigma^j_\infty \widetilde{\psi}^j|^{\frac{4}{d-2}} \nabla \sigma^j_\infty \widetilde{\psi}^j - |\sigma^j_n \widetilde{\psi}^j|^{\frac{4}{d-2}} \nabla \sigma^j_n \widetilde{\psi}^j \}\big\|_{L^{\frac{2(d+2)}{d+4}}_{t,x}(I)} \\
&\quad + \lim_{n\to\infty} \big\| |\sigma^j_\infty \widetilde{\psi}^j|^{\frac{8-2d}{d-2}} (\sigma^j_\infty \widetilde{\psi}^j)^2 \overline{\nabla \sigma^j_\infty \widetilde{\psi}^j} - |\sigma^j_n \widetilde{\psi}^j|^{\frac{8-2d}{d-2}} (\sigma^j_n \widetilde{\psi}^j)^2 \overline{\nabla \sigma^j_n \widetilde{\psi}^j} \}\big\|_{L^{\frac{2(d+2)}{d+4}}_{t,x}(I)} \\
&\lesssim \lim_{n\to\infty} \{\|\sigma^j_\infty \widetilde{\psi}^j\|^{\frac{6-d}{d-2}}_{W_{2^*}} + \|\sigma^j_n \widetilde{\psi}^j\|^{\frac{6-d}{d-2}}_{W_{2^*}(I)}\} \|\sigma^j_\infty \widetilde{\psi}^j - \sigma^j_n \widetilde{\psi}^j\|_{W_{2^*}(I)} \|\langle\nabla\rangle \widetilde{\psi}^j\|_{V_{2+\frac{4}{d}}(I)} \\
&\quad + \lim_{n\to\infty} \|\sigma^j_n \widetilde{\psi}^j\|^{\frac{4}{d-2}}_{W_{2^*}(I)} \|\nabla \{\sigma^j_\infty \widetilde{\psi}^j - \sigma^j_\infty \widetilde{\psi}^j\}\|_{V_{2+\frac{4}{d}}(I)} \\
&= 0.
\end{aligned}
$$

Next, we consider the second term on the right-hand side of (8.94). We see from the Mihlin multiplier theorem, the assumption (8.91) and Lemma 8.0.10 that

(8.98)
$$\begin{aligned}&\lim_{n\to\infty}\lambda_n^j\big\|\{|(\lambda_n^j)^{-1}\nabla|-\langle(\lambda_n^j)^{-1}\nabla\rangle\}F^{\ddagger}[\sigma_n^j\widetilde{\psi}^j]\big\|_{L_{t,x}^{\frac{2(d+2)}{d+4}}(I)}\\&\lesssim\lim_{n\to\infty}\lambda_n^j\big\|F^{\ddagger}[\sigma_n^j\widetilde{\psi}^j]\big\|_{L_{t,x}^{\frac{2(d+2)}{d+4}}(I)}\lesssim\lim_{n\to\infty}\big\|\sigma_n^j\widetilde{\psi}^j\big\|_{W_{2^*}(I)}^{\frac{4}{d-2}}\|\lambda_n^j\sigma_n^j\widetilde{\psi}^j\|_{V_{2+\frac{4}{d}}(I)}=0.\end{aligned}$$

Putting (8.94) through (8.98) together, we find that

$$\lim_{n\to\infty}\big\|\langle\nabla\rangle(\sigma_\infty^j)^{-1}F^{\ddagger}[G_n^j\sigma_\infty^j\widetilde{\psi}^j]-\langle\nabla\rangle F^{\ddagger}[G_n^j\sigma_n^j\widetilde{\psi}^j]\big\|_{L_{t,x}^{\frac{2(d+2)}{d+4}}(I_n^j)}=0. \tag{8.99}$$

We can also verify in a way similar to the estimates (8.94) and (8.98) that

(8.100)
$$\begin{aligned}&\lim_{n\to\infty}\big\|\langle\nabla\rangle F^{\dagger}[G_n^j\sigma_n^j\widetilde{\psi}^j]\big\|_{L_{t,x}^{\frac{2(d+2)}{d+4}}(I_n^j)}\\&=\lim_{n\to\infty}(\lambda_n^j)^{\frac{4-(d-2)(p-1)}{2}}\big\|\lambda_n^j\langle(\lambda_n^j)^{-1}\nabla\rangle F^{\dagger}[\sigma_n^j\widetilde{\psi}^j]\big\|_{L_{t,x}^{\frac{2(d+2)}{d+4}}(I_n^j)}\\&\lesssim\lim_{n\to\infty}(\lambda_n^j)^{\frac{4-(d-2)(p-1)}{2}}\lambda_n^j\big\|\{|(\lambda_n^j)^{-1}\nabla|-\langle(\lambda_n^j)^{-1}\nabla\rangle\}F^{\dagger}[\sigma_n^j\widetilde{\psi}^j]\big\|_{L_{t,x}^{\frac{2(d+2)}{d+4}}(I_n^j)}\\&\quad+\lim_{n\to\infty}(\lambda_n^j)^{\frac{4-(d-2)(p-1)}{2}}\big\|\nabla F^{\dagger}[\sigma_n^j\widetilde{\psi}^j]\big\|_{L_{t,x}^{\frac{2(d+2)}{d+4}}(I_n^j)}\\&\lesssim\lim_{n\to\infty}(\lambda_n^j)^{\frac{4-(d-2)(p-1)}{2}}\lambda_n^j\big\|F^{\dagger}[\sigma_n^j\widetilde{\psi}^j]\big\|_{L_{t,x}^{\frac{2(d+2)}{d+4}}(I_n^j)}\\&\quad+\lim_{n\to\infty}(\lambda_n^j)^{\frac{4-(d-2)(p-1)}{2}}\big\|\nabla F^{\dagger}[\sigma_n^j\widetilde{\psi}^j]\big\|_{L_{t,x}^{\frac{2(d+2)}{d+4}}(I_n^j)}\\&\lesssim\lim_{n\to\infty}(\lambda_n^j)^{\frac{4-(d-2)(p-1)}{2}}\big\|\sigma_n^j\widetilde{\psi}^j\big\|_{W_{p+1}(I)}^{p-1}\|\lambda_n^j\sigma_n^j\widetilde{\psi}^j\|_{V_{2+\frac{4}{d}}(I)}\\&\quad+\lim_{n\to\infty}(\lambda_n^j)^{\frac{4-(d-2)(p-1)}{2}}\big\|\sigma_n^j\widetilde{\psi}^j\big\|_{W_{p+1}(I)}^{p-1}\|\nabla\sigma_n^j\widetilde{\psi}^j\|_{V_{2+\frac{4}{d}}(I)}\\&\le\lim_{n\to\infty}\|\lambda_n^j\sigma_n^j\widetilde{\psi}^j\|_{V_{2+\frac{4}{d}}(I)}^{(p-1)(1-s_p)}\|\sigma_n^j\widetilde{\psi}^j\|_{W_{2^*}(I)}^{(p-1)s_p}\|\lambda_n^j\sigma_n^j\widetilde{\psi}^j\|_{V_{2+\frac{4}{d}}(I)}\\&\quad+\lim_{n\to\infty}\|\lambda_n^j\sigma_n^j\widetilde{\psi}^j\|_{V_{2+\frac{4}{d}}(I)}^{(p-1)(1-s_p)}\|\lambda_n^j\sigma_n^j\widetilde{\psi}^j\|_{W_{2^*}(I)}^{(p-1)s_p}\|\nabla\sigma_n^j\widetilde{\psi}^j\|_{V_{2+\frac{4}{d}}(I)}\\&=0.\end{aligned}$$

Then, the desired result (8.92) follows from (8.99) and (8.100). □

LEMMA 8.0.12. *For any $\delta>0$ and any $k\ge 1$, there exists a number $N(\delta,k)$ such that for any $n\ge N(\delta,k)$,*

$$\sum_{\substack{1\le j\le k\\ \lambda_\infty^j=\infty}}\|\langle\nabla\rangle F[\psi_n^j]\|_{L_{t,x}^{\frac{2(d+2)}{d+4}}(\mathbb{R})}\le\delta, \tag{8.101}$$

where the sum is taken over all integers j satisfying $1 \le j \le k$ and $\lambda_\infty^j = \infty$.

Proof of Lemma 8.0.12. Let q denote p or $2^* - 1$, and let $j \ge 1$ be a number for which $\lambda_\infty^j = \infty$, so that $\psi_n^j = G_n^j \sigma_n^j e^{it\Delta}\widetilde{u}^j$. Then, we see from Hölder's inequality, Sobolev's embedding, Strichartz' estimate, Lemma 8.0.7 with $\sigma_\infty^j = 0$ (see (8.56)) that

$$\begin{aligned}
&\lim_{n\to\infty} \||\psi_n^j|^{q-1}\psi_n^j\|_{L_{t,x}^{\frac{2(d+2)}{d+4}}(\mathbb{R})} \\
&\le \lim_{n\to\infty} \|\psi_n^j\|_{W_{q+1}(I_n)}^{q-1} \|\psi_n^j\|_{V_{2+\frac{4}{d}}(\mathbb{R})} \\
&\lesssim \lim_{n\to\infty} (\lambda_n^j)^{(q-1)(1-s_q)} \||\nabla|^{s_q}\sigma_n^j e^{it\Delta}\widetilde{u}^j\|_{V_{q+1}(I_n)}^{q-1} \lambda_n^j \|\sigma_n^j e^{it\Delta}\widetilde{u}^j\|_{V_{2+\frac{4}{d}}(\mathbb{R})} \\
&\lesssim \lim_{n\to\infty} (\lambda_n^j)^{(q-1)(1-s_q)} \||\nabla|^{s_q}\sigma_n^j \widetilde{u}^j\|_{L^2}^{q-1} \|\lambda_n^j \sigma_n^j \widetilde{u}^j\|_{L^2} \\
&\lesssim \lim_{n\to\infty} \|\lambda_n^j \sigma_n^j \widetilde{u}^j\|_{L^2}^{(q-1)(1-s_q)} \|\nabla \sigma_n^j \widetilde{u}^j\|_{L^2}^{(q-1)s_q} \|\lambda_n^j \sigma_n^j \widetilde{u}^j\|_{L^2} = 0.
\end{aligned} \tag{8.102}$$

Similarly, we can verify that

$$\begin{aligned}
\lim_{n\to\infty} \|\nabla\{|\psi_n^j|^{q-1}\psi_n^j\}\|_{L_{t,x}^{\frac{2(d+2)}{d+4}}(\mathbb{R})} &\lesssim \lim_{n\to\infty} \||\psi_n^j|^{q-1}|\nabla\psi_n^j|\|_{L_{t,x}^{\frac{2(d+2)}{d+4}}(\mathbb{R})} \\
&\le \lim_{n\to\infty} \|\psi_n^j\|_{W_{q+1}(I_n)}^{q-1} \|\nabla\psi_n^j\|_{V_{2+\frac{4}{d}}(\mathbb{R})} = 0.
\end{aligned} \tag{8.103}$$

Thus, we find from (8.102) and (8.103) that for any $\delta > 0$ and any $k \ge 1$, there exists $N(\delta, k) \ge 1$ such that for any $n \ge N(\delta, k)$,

$$\sum_{\substack{1\le j\le k \\ \lambda_\infty^j = \infty}} \|\langle\nabla\rangle\{|\psi_n^j|^{q-2}\psi_n^j\}\|_{L_{t,x}^{\frac{2(d+2)}{d+4}}(\mathbb{R})} \le \delta, \tag{8.104}$$

which gives us the desired result. □

Lemma 8.0.13. *Let $j \ge 1$, and let I be an interval on which $\widetilde{\psi}^j$ exists. Then, we have*

$$\sup_{n\ge1} \|\langle\nabla\rangle\psi_n^j\|_{St(I_n^j)} \lesssim \|\langle\nabla\rangle\widetilde{\psi}^j\|_{St(I)}, \tag{8.105}$$

where $I_n^j := ((\lambda_n^j)^2 \inf I + t_n^j,\ (\lambda_n^j)^2 \sup I + t_n^j)$.

Proof of Lemma 8.0.13. We can easily verify that

$$\|\langle\nabla\rangle\psi_n^j\|_{L_t^\infty L_x^2(I_n^j)} \le \|\lambda_n^j\sigma_n^j\widetilde{\psi}^j\|_{L_t^\infty L_x^2(I)} + \|\nabla\sigma_n^j\widetilde{\psi}^j\|_{L_t^\infty L_x^2(I)}, \tag{8.106}$$

$$\|\langle\nabla\rangle\psi_n^j\|_{L_t^2 L_x^{2^*}(I_n^j)} \le \|\lambda_n^j\sigma_n^j\widetilde{\psi}^j\|_{L_t^2 L_x^{2^*}(I)} + \|\nabla\sigma_n^j\widetilde{\psi}^j\|_{L_t^2 L_x^{2^*}(I)}. \tag{8.107}$$

When $\lambda_\infty^j \in \{0, 1\}$, these estimates (8.106) and (8.107) together with Lemma 8.0.8 prove the lemma. Assume $\lambda_\infty^j = \infty$, so that $\widetilde{\psi}^j(t) = e^{it\Delta}\widetilde{u}^j$. Then, we see from (8.106), (8.107) and Strichartz' estimate that

$$\begin{aligned}
\|\langle\nabla\rangle\psi_n^j\|_{L_t^\infty L_x^2(I_n^j)} + \|\langle\nabla\rangle\psi_n^j\|_{L_t^2 L_x^{2^*}(I_n^j)} &\lesssim \|\lambda_n^j\sigma_n^j\widetilde{u}^j\|_{L^2} + \|\nabla\sigma_n^j\widetilde{u}^j\|_{L^2} \\
&\lesssim \|\langle\nabla\rangle\widetilde{u}^j\|_{L^2} = \|\langle\nabla\rangle\widetilde{\psi}^j\|_{L_t^\infty L_x^2(I)}.
\end{aligned} \tag{8.108}$$

Thus, we have completed the proof. □

LEMMA 8.0.14 (cf. Lemma 6.8 in [**3**]). *There exists a number J_0 such that $I^j_{\max} = \mathbb{R}$ for any $j > J_0$; and for any $r \geq 2$,*

$$\sum_{j>J_0} \|\langle \nabla \rangle \widetilde{\psi}^j\|^r_{St(\mathbb{R})} \lesssim \sum_{j>J_0} \|\widetilde{u}^j\|^r_{H^1} < \infty. \tag{8.109}$$

PROOF OF LEMMA 8.0.14. It follows from (8.49) and the uniform boundedness (8.38) that

$$\lim_{n\to\infty} \sum_{j=1}^{k} \|g^j_n \sigma^j_n e^{-i\frac{t^j_n}{(\lambda^j_n)^2}\Delta} \widetilde{u}^j\|^2_{H^1} \lesssim 1 \tag{8.110}$$

for all $k \geq 1$. Furthermore, we see from Lemma 8.0.7 that: if $\lambda^j_\infty = 0$, then

$$\lim_{n\to\infty} \|g^j_n \sigma^j_n e^{-i\frac{t^j_n}{(\lambda^j_n)^2}\Delta} \widetilde{u}^j\|_{L^2} = \lim_{n\to\infty} \|\lambda^j_n \sigma^j_n \widetilde{u}^j\|_{L^2} = 0, \tag{8.111}$$

$$\lim_{n\to\infty} \|\nabla g^j_n \sigma^j_n e^{-i\frac{t^j_n}{(\lambda^j_n)^2}\Delta} \widetilde{u}^j\|_{L^2} = \lim_{n\to\infty} \|\nabla \sigma^j_n \widetilde{u}^j\|_{L^2} = \|\widetilde{u}^j\|_{H^1}; \tag{8.112}$$

and if $\lambda^j_\infty \in \{1, \infty\}$, then

$$\lim_{n\to\infty} \|g^j_n \sigma^j_n e^{-i\frac{t^j_n}{(\lambda^j_n)^2}\Delta} \widetilde{u}^j\|_{H^1} = \|\widetilde{u}^j\|_{H^1}. \tag{8.113}$$

Putting (8.110) through (8.113) together, we obtain that

$$\sum_{j=1}^{\infty} \|\widetilde{u}^j\|^2_{H^1} = \lim_{k\to\infty} \sum_{j=1}^{k} \|\widetilde{u}^j\|^2_{H^1} \lesssim \lim_{k\to\infty} \sum_{j=1}^{k} \lim_{n\to\infty} \|g^j_n \sigma^j_n e^{-i\frac{t^j_n}{(\lambda^j_n)^2}\Delta} \widetilde{u}^j\|^2_{H^1} \lesssim 1. \tag{8.114}$$

In particular, for any $\delta > 0$, we can take $J(\delta) \geq 1$ such that for any $j \geq J(\delta)$,

$$\|\widetilde{u}^j\|_{H^1} < \delta. \tag{8.115}$$

Note here that if $\lambda^j_\infty = 0$, then (8.115) is rewritten as $\|\nabla \sigma^j_\infty \widetilde{u}^j\|^2_{L^2} < \delta$. Since $\widetilde{\psi}^j$ satisfies (8.74) and (8.76), the small-data theory (Lemma D.0.1) together with (8.115) implies that there exists a number J_0 such that for any $j \geq J_0$,

$$\|\langle \nabla \rangle \widetilde{\psi}^j\|_{St(\mathbb{R})} \lesssim \|\widetilde{u}^j\|_{H^1}. \tag{8.116}$$

Then, the claim (8.109) follows from (8.114), (8.116) and Jensen's inequality. □

LEMMA 8.0.15. *For any number $k \geq 1$, there exists a number $N(k)$ such that for any $n \geq N(k)$,*

$$\sum_{j=1}^{k} \mathcal{I}_\omega\big(g^j_n \sigma^j_n e^{-i\frac{t^j_n}{(\lambda^j_n)^2}\Delta} \widetilde{u}^j\big) + \mathcal{I}_\omega(w^k_n) \leq m_\omega - \frac{\kappa_1(R_*)}{10d(p-1)}. \tag{8.117}$$

PROOF OF LEMMA 8.0.15. It follows from (8.32), (8.37) and $\mathcal{S}_\omega(\psi_n) \leq m_\omega + \varepsilon_n$ that for any numbers n and k,

$$\begin{aligned} m_\omega - \frac{\kappa_1(R_*)}{2d(p-1)} &\geq m_\omega + \varepsilon_n - \frac{2}{d(p-1)}\mathcal{K}(\psi_n(0)) \\ &\geq \mathcal{S}_\omega(\psi_n) - \frac{2}{d(p-1)}\mathcal{K}(\psi_n(0)) = \mathcal{I}_\omega(\psi_n(0)). \end{aligned} \tag{8.118}$$

This together with (8.53) gives us the desired result (8.117). □

LEMMA 8.0.16. *For any number $k \ge 1$, there exists a number $N(k)$ such that for any $n \ge N(k)$ and any $j \le k$ for which the linear profile $\widetilde{u}^j$ is non-trivial,*

$$\mathcal{H}\big(g_n^j \sigma_n^j e^{-i\frac{t_n^j}{(\lambda_n^j)^2}\Delta}\widetilde{u}^j\big) > \frac{1}{2}\mathcal{K}\big(g_n^j \sigma_n^j e^{-i\frac{t_n^j}{(\lambda_n^j)^2}\Delta}\widetilde{u}^j\big) > 0, \tag{8.119}$$

$$\mathcal{H}(w_n^k) \ge \frac{1}{2}\mathcal{K}(w_n^k) \ge 0. \tag{8.120}$$

PROOF OF LEMMA 8.0.16. Since $\mathcal{I}_\omega$ is non-negative, Lemma 8.0.15 together with (1.10) shows that for any $k \ge 1$, there exists a number $N(k)$ such that for any $n \ge N(k)$ and any $j \le k$ for which $\widetilde{u}^j$ is non-trivial,

$$\mathcal{K}\big(g_n^j \sigma_n^j e^{-i\frac{t_n^j}{(\lambda_n^j)^2}\Delta}\widetilde{u}^j\big) > 0, \quad \mathcal{K}(w_n^k) \ge 0. \tag{8.121}$$

This together with (1.6) gives us the desired result. □

LEMMA 8.0.17. *There is at most one number j_0 such that $\lambda_\infty^{j_0} \in \{0, 1\}$ and*

$$\begin{cases} \mathcal{S}_\omega(\sigma_\infty^{j_0}\widetilde{\psi}^{j_0}) \ge m_\omega & \text{if } \lambda_\infty^{j_0} = 1, \\ \mathcal{H}^\ddagger(\sigma_\infty^{j_0}\widetilde{\psi}^{j_0}) \ge \mathcal{H}^\ddagger(W) & \text{if } \lambda_\infty^{j_0} = 0. \end{cases} \tag{8.122}$$

PROOF OF LEMMA 8.0.17. First, we consider a number k such that

$$\mathcal{S}_\omega\big(g_n^k \sigma_n^k e^{-i\frac{t_n^k}{(\lambda_n^k)^2}\Delta}\widetilde{u}^k\big) \le \frac{2}{3}m_\omega. \tag{8.123}$$

Then, it follows from Lemma 8.0.7, Lemma 8.0.9, (8.74), (8.79), (1.34) and (1.35) that for any sufficiently large number n,

$$\begin{cases} \mathcal{S}_\omega(\sigma_\infty^k \widetilde{\psi}^k) = \mathcal{S}_\omega\Big(\sigma_\infty^k \widetilde{\psi}^k\Big(-\frac{t_n^k}{(\lambda_n^k)^2}\Big)\Big) < m_\omega & \text{if } \lambda_\infty^k = 1, \\ \mathcal{H}^\ddagger(\sigma_\infty^k \widetilde{\psi}^k) = \mathcal{H}^\ddagger\Big(\sigma_\infty^k \widetilde{\psi}^k\Big(-\frac{t_n^k}{(\lambda_n^k)^2}\Big)\Big) < \mathcal{H}^\ddagger(W) & \text{if } \lambda_\infty^k = 0. \end{cases} \tag{8.124}$$

Thus, it suffices for the desired result to show that there is at most one number j_0 such that for any sufficiently large n,

$$\mathcal{S}_\omega\big(g_n^{j_0} \sigma_n^{j_0} e^{-i\frac{t_n^{j_0}}{(\lambda_n^{j_0})^2}\Delta}\widetilde{u}^{j_0}\big) \ge \frac{2}{3}m_\omega. \tag{8.125}$$

We see from Lemma 8.0.16 that for any $j \ge 1$ and any sufficiently large n,

$$\mathcal{S}_\omega\big(g_n^j \sigma_n^j e^{-i\frac{t_n^j}{(\lambda_n^j)^2}\Delta}\widetilde{u}^j\big) \ge 0. \tag{8.126}$$

Hence, it follows from (8.52) and $\mathcal{S}_\omega(\psi_n(0)) < m_\omega + \varepsilon_n$ that there are at most one linear profile satisfying (8.125). □

Now, using Lemma 8.0.17 and reordering the indices, we may assume that for any $j \ge 2$,

$$\begin{cases} \mathcal{S}_\omega(\sigma_\infty^j \widetilde{\psi}^j) < m_\omega & \text{if } \lambda_\infty^j = 1, \\ \mathcal{H}^\ddagger(\sigma_\infty^j \widetilde{\psi}^j) < \mathcal{H}^\ddagger(W) & \text{if } \lambda_\infty^j = 0. \end{cases} \tag{8.127}$$

LEMMA 8.0.18. *Assume (8.127). Then, for any $j \geq 1$ for which the nonlinear profile $\widetilde{\psi}^j$ is non-trivial and $\lambda_\infty^j \in \{0,1\}$, there exists a number $N(j)$ such that for any $n \geq N(j)$,*

$$\begin{cases} \mathcal{K}\big(\widetilde{\psi}^j\big(-\frac{t_n^j}{(\lambda_n^j)^2}\big)\big) > 0 & \text{if } \lambda_\infty^j = 1, \\ \mathcal{K}^\ddagger\big(\sigma_\infty^j \widetilde{\psi}^j\big(-\frac{t_n^j}{(\lambda_n^j)^2}\big) > 0 & \text{if } \lambda_\infty^j = 0. \end{cases} \tag{8.128}$$

Furthermore, if $\lambda_\infty^j = 0$ and $\mathcal{H}^\ddagger(\sigma_\infty^j \widetilde{\psi}^j) < \mathcal{H}^\ddagger(W)$, then for any $n \geq N(j)$,

$$\big\|\nabla \sigma_\infty^j \widetilde{\psi}^j\big(-\frac{t_n^j}{(\lambda_n^j)^2}\big)\big\|_{L^2}^2 < \|\nabla W\|_{L^2}^2. \tag{8.129}$$

PROOF OF LEMMA 8.0.18. We see from Lemma 8.0.15 that there exists a number $N(j)$ such that for any $n \geq N(j)$,

$$\mathcal{I}_\omega\big(g_n^j \sigma_n^j e^{-i\frac{t_n^j}{(\lambda_n^j)^2}\Delta} \widetilde{u}^j\big) \leq m_\omega - \frac{\kappa_1(R_*)}{100d(p-1)}. \tag{8.130}$$

When $\lambda_\infty^j = 1$, the desired result (8.128) follows from (8.79), (8.130) and (1.10). In order to prove (8.128) in the case $\lambda_\infty^j = 0$, note that it follows from (1.30) and (8.130) that for any $n \geq N(j)$,

$$\begin{aligned} &\mathcal{I}_\omega^\ddagger\big(g_n^j \sigma_n^j e^{-i\frac{t_n^j}{(\lambda_n^j)^2}\Delta} \widetilde{u}^j\big) = \mathcal{H}^\ddagger\big(g_n^j \sigma_n^j e^{-i\frac{t_n^j}{(\lambda_n^j)^2}\Delta} \widetilde{u}^j\big) - \frac{2}{d(p-1)} \mathcal{K}^\ddagger\big(g_n^j \sigma_n^j e^{-i\frac{t_n^j}{(\lambda_n^j)^2}\Delta} \widetilde{u}^j\big) \\ &\leq \mathcal{S}_\omega\big(g_n^j \sigma_n^j e^{-i\frac{t_n^j}{(\lambda_n^j)^2}\Delta} \widetilde{u}^j\big) - \frac{2}{d(p-1)} \mathcal{K}\big(g_n^j \sigma_n^j e^{-i\frac{t_n^j}{(\lambda_n^j)^2}\Delta} \widetilde{u}^j\big) \\ &= \mathcal{I}_\omega\big(g_n^j \sigma_n^j e^{-i\frac{t_n^j}{(\lambda_n^j)^2}\Delta} \widetilde{u}^j\big) \leq m_\omega - \frac{\kappa_1(R_*)}{100d(p-1)}. \end{aligned} \tag{8.131}$$

Furthermore, we see from (8.131) and (8.79) that for any $n \geq N(j)$,

$$\mathcal{I}_\omega^\ddagger\big(\sigma_\infty^j \widetilde{\psi}^j\big(-\frac{t_n^j}{(\lambda_n^j)^2}\big)\big) = \lim_{n\to\infty} \mathcal{I}^\ddagger\big(g_n^j \sigma_n^j e^{-i\frac{t_n^j}{(\lambda_n^j)^2}\Delta} \widetilde{u}^j\big) \leq m_\omega - \frac{\kappa_1(R_*)}{100d(p-1)}. \tag{8.132}$$

Thus, when $\lambda_\infty^j = 0$, (1.29) together with (1.34), (1.35) and (8.132) shows that for any $n \geq N(j)$,

$$\mathcal{K}^\ddagger\big(\sigma_\infty^j \widetilde{\psi}^j\big(-\frac{t_n^j}{(\lambda_n^j)^2}\big)\big) = \big\|\nabla \sigma_\infty^j \widetilde{\psi}^j\big(-\frac{t_n^j}{(\lambda_n^j)^2}\big)\big\|_{L^2}^2 - \big\|\sigma_\infty^j \widetilde{\psi}^j\big(-\frac{t_n^j}{(\lambda_n^j)^2}\big)\big\|_{L^{2^*}}^{2^*} > 0, \tag{8.133}$$

which completes the proof of (8.128).

Next, we shall show (8.129). Assume $\lambda_\infty^j = 0$ and $\mathcal{H}^\ddagger(\sigma_\infty^j \widetilde{\psi}^j) < \mathcal{H}^\ddagger(W)$. Then, we see from (8.133), (1.34) and (1.28) that for any $n \geq N(j)$,

$$
\begin{aligned}
&\frac{1}{d}\|\nabla \sigma_\infty^j \widetilde{\psi}^j\Big(-\frac{t_n^j}{(\lambda_n^j)^2}\Big)\|_{L^2}^2 \\
&= \frac{1}{2}\|\nabla \sigma_\infty^j \widetilde{\psi}^j\Big(-\frac{t_n^j}{(\lambda_n^j)^2}\Big)\|_{L^2}^2 - \frac{1}{2^*}\|\nabla \sigma_\infty^j \widetilde{\psi}^j\Big(-\frac{t_n^j}{(\lambda_n^j)^2}\Big)\|_{L^2}^2 \\
&\leq \frac{1}{2}\|\nabla \sigma_\infty^j \widetilde{\psi}^j\Big(-\frac{t_n^j}{(\lambda_n^j)^2}\Big)\|_{L^2}^2 - \frac{1}{2^*}\|\sigma_\infty^j \widetilde{\psi}^j\Big(-\frac{t_n^j}{(\lambda_n^j)^2}\Big)\|_{L^{2^*}}^{2^*} \\
&= \mathcal{H}^\ddagger(\sigma_\infty^j \widetilde{\psi}^j) < \mathcal{H}^\ddagger(W) = \frac{1}{d}\|\nabla W\|_{L^2}^2.
\end{aligned} \tag{8.134}
$$

Thus, we have completed the proof. □

We see from Lemma 8.0.18 and (8.127) that for any $j \geq 2$ for which $\widetilde{\psi}^j$ is non-trivial and $\lambda_\infty^j \neq \infty$,

$$
\begin{cases} \sigma_\infty^j \widetilde{\psi}^j \in PW_{\omega,+} & \text{if} \quad \lambda_\infty^j = 1, \\ \sigma_\infty^j \widetilde{\psi}^j \in PW_+^\ddagger & \text{if} \quad \lambda_\infty^j = 0. \end{cases} \tag{8.135}
$$

Furthermore, since $\widetilde{\psi}^j(t) = e^{it\Delta}\widetilde{u}^j$ for $j \geq 1$ with $\lambda_\infty^j = \infty$, it follows from Theorem 1.0.2, Theorem 1.0.3 and Lemma 8.0.14 that $I_{\max}^j = \mathbb{R}$ for any $j \geq 2$, and

$$
\sup_{j \geq 2} \|\langle \nabla \rangle \widetilde{\psi}^j\|_{St(\mathbb{R})} < \infty. \tag{8.136}
$$

8.0.2.4. *Existence of bad profile.* In the previous section, we showed that there are at most one bad profile (see Lemma 8.0.17), and the candidate is $\widetilde{\psi}^1$ (see (8.136)). Our aim in this section is to show that for any $T \in (T_{\min}^1, T_{\max}^1)$,

$$
\|\sigma_\infty^1 \widetilde{\psi}^1\|_{\mathcal{W}^1([T, T_{\max}^1))} = \infty, \tag{8.137}
$$

where $\mathcal{W}^1$ is the function space defined by (8.73) (see Proposition 8.0.25 below). To this end, we observe properties of $\widetilde{\psi}^1$ on an interval where

$$
\|\sigma_\infty^1 \widetilde{\psi}^1\|_{\mathcal{W}^1(I)} < \infty. \tag{8.138}
$$

LEMMA 8.0.19. *For any interval I satisfying* (8.138), *we can take a constant $A(I) > 0$ such that*

$$
\|\langle \nabla \rangle \widetilde{\psi}^1\|_{St(I)} \leq A(I). \tag{8.139}
$$

PROOF OF LEMMA 8.0.19. When $\lambda_\infty^1 = \infty$, $\widetilde{\psi}^1(t) = e^{it\Delta}\widetilde{u}^1$ and therefore the claim follows from Strichartz' estimate. Moreover, when I is a compact interval in $I_{\max}^1$, the claim follows from the well-posedness theory. Thus, we may assume that $\lambda_\infty^j \in \{0, 1\}$, and $\sup I = T_{\max}^1$ or $\inf I = T_{\min}^1$. We only consider the case where $\sup I = T_{\max}^1$ and $T_{\min}^1 < \inf I$. The same proof is applicable for the other cases. Since $\sup I = T_{\max}^1$, the blowup criterion (cf. Theorem 4.1 in [**3**] and Lemma 2.11 in [**18**]) together with (8.138) shows that $T_{\max}^1 = \infty$ and

$$
\|\sigma_\infty^1 \widetilde{\psi}^1\|_{\mathcal{W}^1([\inf I, \infty))} < \infty. \tag{8.140}
$$

We shall show that

$$\sup_{t\in[\inf I,\infty)} \|\widetilde{\psi}^1(t)\|_{H^1} < \infty. \tag{8.141}$$

Suppose for contradiction that (8.141) was false. Then, we could take a sequence $\{t_n\}$ in $[\inf I, \infty)$ such that $\lim_{n\to\infty} t_n = \infty$ and

$$\lim_{n\to\infty} \left\|\widetilde{\psi}^1(t_n)\right\|_{H^1} = \infty. \tag{8.142}$$

Assume $\lambda_\infty^1 = 1$. Then, $\widetilde{\psi}^1$ is a solution to (NLS), and we see from Strichartz' estimate and Hölder's inequality that for any $t_m < t_n$,

$$\begin{aligned}&\left\|e^{-it_m\Delta}\widetilde{\psi}^1(t_m) - e^{-it_n\Delta}\widetilde{\psi}^1(t_n)\right\|_{H^1} \le \sup_{T\in[t_m,t_n]} \left\| \int_{t_m}^T e^{-it\Delta} F[\widetilde{\psi}^1(t)]\,dt\right\|_{H^1}\\ &\lesssim \|\widetilde{\psi}^1\|_{W_{p+1}([t_m,\infty))}^{p-1} \|\langle\nabla\rangle\widetilde{\psi}^1\|_{V_{p+1}([t_m,\infty))} + \|\widetilde{\psi}^1\|_{W_{2^*}([t_m,\infty))}^{\frac{4}{d-2}} \|\langle\nabla\rangle\widetilde{\psi}^1\|_{V_{2^*}([t_m,\infty))}.\end{aligned} \tag{8.143}$$

Moreover, we see from (8.140) and $\sigma_\infty^1 = 1$ that

$$\lim_{T\to\infty} \|\widetilde{\psi}^1\|_{W_{p+1}([T,\infty))\cap W_{2^*}([T,\infty))} = 0. \tag{8.144}$$

In particular, for any $\delta > 0$, there exists $T(\delta) > 0$ such that

$$\|\widetilde{\psi}^1\|_{W_{p+1}([T(\delta),\infty))\cap W_{2^*}([T(\delta),\infty))} \le \delta. \tag{8.145}$$

An estimate similar to (8.143) together with (8.145) also yields that

$$\begin{aligned}&\|\langle\nabla\rangle\widetilde{\psi}^1\|_{V_{p+1}([T(\delta),\infty))\cap V_{2^*}([T(\delta),\infty))}\\ &\lesssim \|\widetilde{\psi}^1(T(\delta))\|_{H^1} + (\delta^{p-1} + \delta^{\frac{4}{d-2}})\|\langle\nabla\rangle\widetilde{\psi}^1\|_{V_{p+1}([T(\delta),\infty))\cap V_{2^*}([T(\delta),\infty))}.\end{aligned} \tag{8.146}$$

Thus, we can take $\delta_0 > 0$ such that

$$\|\langle\nabla\rangle\widetilde{\psi}^1\|_{V_{p+1}([T(\delta_0),\infty))\cap V_{2^*}([T(\delta_0),\infty))} \lesssim \|\widetilde{\psi}^1(T(\delta_0))\|_{H^1}. \tag{8.147}$$

Combining (8.143) with (8.144) and (8.147), we find that $\{e^{-it_n\Delta}\widetilde{\psi}^1(t_n)\}$ is a Cauchy sequence in $H^1(\mathbb{R}^d)$. Similarly, when $\lambda_\infty^1 = 0$, we can verify $\{e^{-it_n\Delta}\sigma_\infty^1\widetilde{\psi}^1(t_n)\}$ is a Cauchy sequence in $\dot{H}^1(\mathbb{R}^d)$. Thus, in both cases $\lambda_\infty^1 = 1$ and $\lambda_\infty^1 = 0$, we can take $\phi_+ \in H^1(\mathbb{R}^d)$ such that

$$\lim_{n\to\infty} \|\widetilde{\psi}^1(t_n)\|_{H^1} = \|\phi_+\|_{H^1} < \infty. \tag{8.148}$$

However, this contradicts (8.142). Hence, we have proved (8.141).

Now, we are able to show (8.139). If $\lambda_\infty^1 = 1$, then $e[\widetilde{\psi}^1] \equiv 0$ (see (1.66) for the definition of $e[\widetilde{\psi}^1]$), and Lemma D.0.2 together with (8.140) and (8.141) gives us the desired estimate (8.139). If $\lambda_\infty^1 = 0$, then $e^\ddagger[\sigma_\infty^1\widetilde{\psi}^1] \equiv 0$. Moreover, it follows from (8.73), (8.140) and (8.141) that

$$\|\sigma_\infty^1\widetilde{\psi}^1\|_{W_{2^*}([\inf I,\infty))} < \infty, \qquad \sup_{t\in[\inf I,\infty)} \|\nabla\sigma_\infty^1\widetilde{\psi}^1\|_{L^2} < \infty. \tag{8.149}$$

Hence, we see from Lemma D.0.2 that

$$\|\langle\nabla\rangle\widetilde{\psi}^1\|_{St([\inf I,\infty))} = \|\nabla\sigma_\infty^1\widetilde{\psi}^1\|_{St([\inf I,\infty))} < \infty, \tag{8.150}$$

which completes the proof. □

Since the estimate (8.42) is insufficient to control the remainder $e^{it\Delta}w_n^j$ in the Strichartz space $\langle\nabla\rangle^{-1}St(\mathbb{R})$, we need the following estimate:

LEMMA 8.0.20. *Assume* (8.127), *and let q denote p or 2^*-1. Then, for any $\delta>0$, any interval I satisfying* (8.138), *and any number $j\ge 1$, there exists a number $K(\delta,I,j)$ with the following property: for any $k\ge K(\delta,I,j)$, there exists a number $N(\delta,I,j,k)$ such that for any $n\ge N(\delta,I,j,k)$ and $s\in\{0,1\}$,*

$$\big\|\psi_n^j|\nabla|^s e^{it\Delta}w_n^k\big\|_{L_{t,x}^{\frac{2(d+2)(q-1)}{d(q-1)+4}}(I_n)}\le\delta, \tag{8.151}$$

where $I_n:=((\lambda_n^1)^2\inf I+t_n^1,\ (\lambda_n^1)^2\sup I+t_n^1)$.

PROOF OF LEMMA 8.0.20. Let I be an interval on which (8.138) holds. Put $I^1:=I$ and $I^j:=\mathbb{R}$ for $j\ge 2$. Then, it follows from Lemma 8.0.19 and (8.136) that there exists a constant $A(I)>0$ such that

$$\sup_{j\ge 1}\|\langle\nabla\rangle\widetilde{\psi}^j\|_{St(I^j)}\le A(I). \tag{8.152}$$

Moreover, we see from (8.80) and elementary computations that for any numbers $j,k,n\ge 1$ and $s\in\{0,1\}$,

$$\begin{aligned}&\|\psi_n^j|\nabla|^s e^{it\Delta}w_n^k\|_{L_{t,x}^{\frac{2(d+2)(q-1)}{d(q-1)+4}}(I_n)}\\&\le(\lambda_n^j)^{2-s_q}\|\sigma_n^j\widetilde{\psi}^j(G_n^j)^{-1}|\nabla|^s e^{it\Delta}w_n^k\|_{L_{t,x}^{\frac{2(d+2)(q-1)}{d(q-1)+4}}(I^j)}.\end{aligned} \tag{8.153}$$

We consider the right-hand side of (8.153) according to s and λ_∞^j:

Case 1.1. Assume $s=0$ and $\lambda_\infty^j\in\{0,1\}$ in (8.153). Then, we see from Hölder's inequality, Lemma 8.0.8, (8.152) and (8.55) that there exists a number $K_1(\delta,I)$ with the following property: for any $k\ge K_1(\delta,I)$, there exists a number $N_1(\delta,I,k)$ such that for any $j\ge 1$ and any $n\ge N_1(\delta,I,k)$,

$$\begin{aligned}&(\lambda_n^j)^{2-s_q}\|\sigma_n^j\widetilde{\psi}^j(G_n^j)^{-1}e^{it\Delta}w_n^k\|_{L_{t,x}^{\frac{2(d+2)(q-1)}{d(q-1)+4}}(I^j)}\\&\le\lambda_n^j\|\sigma_n^j\widetilde{\psi}^j\|_{V_{2+\frac{4}{d}}(I^j)}\|e^{it\Delta}w_n^k\|_{W_{q+1}(\mathbb{R})}\lesssim\|\langle\nabla\rangle\widetilde{\psi}^j\|_{V_{2+\frac{4}{d}}(I^j)}\|e^{it\Delta}w_n^k\|_{W_{q+1}(\mathbb{R})}\le\delta.\end{aligned} \tag{8.154}$$

Case 1.2. Assume $s=0$ and $\lambda_\infty^j=\infty$. Then, $\widetilde{\psi}^j(t)=e^{it\Delta}\widetilde{u}^j$ and we see from an estimate similar to (8.154), Strichartz' estimate, Lemma 8.0.7 and Lemma 8.0.14 that for any $k\ge K_1(\delta,I)$, there exists a number $N_1(\delta,I,k)$ such that for any $j\ge 1$ and any $n\ge N_1(\delta,I,k)$,

$$\begin{aligned}&(\lambda_n^j)^{2-s_q}\|\sigma_n^j\widetilde{\psi}^j(G_n^j)^{-1}e^{it\Delta}w_n^k\|_{L_{t,x}^{\frac{2(d+2)(q-1)}{d(q-1)+4}}(I^j)}\\&\le\lambda_n^j\|e^{it\Delta}\sigma_n^j\widetilde{u}^j\|_{V_{2+\frac{4}{d}}(I^j)}\|e^{it\Delta}w_n^k\|_{W_{q+1}(\mathbb{R})}\lesssim\|\langle\nabla\rangle\widetilde{u}^j\|_{L^2}\|e^{it\Delta}w_n^k\|_{W_{q+1}(\mathbb{R})}\le\delta.\end{aligned} \tag{8.155}$$

Case 2.1. Assume $s=1$ and $\lambda_\infty^j\in\{0,1\}$ in (8.153). Take a sequence $\{v_m^j\}_{m\ge 1}$ of smooth functions on $\mathbb{R}^d\times\mathbb{R}$ with the following properties:

$$\lim_{m\to\infty}\big\|\langle\nabla\rangle(\widetilde{\psi}^j-v_m^j)\big\|_{V_{2+\frac{4}{d}}(I^j)\cap V_{2^*}(I^j)}=0, \tag{8.156}$$

and for each $m \ge 1$, there exist $R^j_m > 0$, $T^j_m > 0$ and $a^j_m \in \mathbb{R}$ such that

$$E^j_m := \{(x,t) \colon |x| \le R^j_m,\ |t - a^j_m| \le T^j_m\} \subset \mathbb{R}^d \times I^j \tag{8.157}$$

and the support of v^j_m is contained in E^j_m.

We see from (8.152) and (8.156) that for each $j \ge 1$, there exists a number $M_1(I,j)$ such that for any $m \ge M_1(I,j)$,

$$\|\langle\nabla\rangle v^j_m\|_{V_{2+\frac{4}{d}}(I^j)\cap V_{2^*}(I^j)} \le A(I) + 1. \tag{8.158}$$

Moreover, it follows from (8.156) that for any $\delta > 0$ and any $j \ge 1$, there exists $M_2(\delta, I, j) \ge 1$ such that for any $m \ge M_2(\delta, I, j)$,

$$\|\langle\nabla\rangle(\widetilde{\psi}^j - v^j_m)\|_{V_{2+\frac{4}{d}}(I^j)\cap V_{2^*}(I^j)} \le \delta. \tag{8.159}$$

Put $m_0 := \max\{M_1(I,j), M_2(\delta, I, j)\}$. We see from the triangle inequality, Hölder's inequality and Sobolev's embedding that the right-hand side of (8.153) with $s = 1$ is estimated as follows:

$$\begin{aligned} &(\lambda^j_n)^{2-s_q}\|\sigma^j_n\widetilde{\psi}^j(G^j_n)^{-1}|\nabla|e^{it\Delta}w^k_n\|_{L^{\frac{2(d+2)(q-1)}{d(q-1)+4}}_{t,x}(I^j)} \\ &\le (\lambda^j_n)^{1-s_q}\|\sigma^j_n\widetilde{\psi}^j - \sigma^j_n v^j_{m_0}\|_{W_{q+1}(I^j)}\||\nabla|e^{it\Delta}w^k_n\|_{V_{2+\frac{4}{d}}(I^j)} \\ &\quad + (\lambda^j_n)^{2-s_q}\|\sigma^j_n v^j_{m_0}(G^j_n)^{-1}|\nabla|e^{it\Delta}w^k_n\|_{L^{\frac{2(d+2)(q-1)}{d(q-1)+4}}_{t,x}(I^j)}. \end{aligned} \tag{8.160}$$

We consider the first term on the right-hand side of (8.160). Using (8.54), Hölder's inequality, Lemma 8.0.8 and (8.159) that if n is sufficiently large dependently on k, then

$$\begin{aligned} &(\lambda^j_n)^{1-s_q}\|\sigma^j_n\widetilde{\psi}^j - \sigma^j_n v^j_{m_0}\|_{W_{q+1}(I^j)}\||\nabla|e^{it\Delta}w^k_n\|_{V_{2+\frac{4}{d}}(I^j)} \\ &\lesssim (\lambda^j_n)^{1-s_q}\|\sigma^j_n\widetilde{\psi}^j - \sigma^j_n v^j_{m_0}\|^{1-s_q}_{V_{2+\frac{4}{d}}(I^j)}\|\sigma^j_n\widetilde{\psi}^j - \sigma^j_n v^j_{m_0}\|^{s_q}_{W_{2^*}(I^j)} \\ &\lesssim \|\langle\nabla\rangle\{\widetilde{\psi}^j - v^j_{m_0}\}\|^{1-s_q}_{V_{2+\frac{4}{d}}(I)}\|\langle\nabla\rangle\{\widetilde{\psi}^j - v^j_{m_0}\}\|^{1-s_q}_{V_{2^*}(I)} \le \delta. \end{aligned} \tag{8.161}$$

Next, we consider the second term on the right-hand side of (8.160). Assume first that $\lambda^j_\infty = 1$, so that $\lambda^j_n \equiv 1$ and $\sigma^j_n \equiv 1$. Then, we see from Hölder's inequality, Sobolev's embedding, (8.158), (8.54), Lemma 2.5 of [**21**] and (8.55) that there exists a number $K_2(\delta, I, j)$ with the following property: for any $k \ge K_2(\delta, I, j)$, there exists $N_2(\delta, I, j, k) \ge 1$ such that for any $n \ge N_2(\delta, I, j, k)$,

$$\begin{aligned}
&(\lambda_n^j)^{2-s_q}\|\sigma_n^j v_{m_0}^j (G_n^j)^{-1}|\nabla|e^{it\Delta}w_n^k\|_{L_{t,x}^{\frac{2(d+2)(q-1)}{d(q-1)+4}}(I^j)} \\
&= \|v_{m_0}^j|\nabla|e^{it\Delta}w_n^k\|_{L^{\frac{2(d+2)(q-1)}{d(q-1)+4}}(E_{m_0}^j)} \\
&\le \|v_{m_0}^j\|_{W_{2^*}(I^j)}\||\nabla|e^{it\Delta}w_n^k\|_{L^{\frac{(d+2)(q-1)}{q+1}}(E_{m_0}^j)} \\
&\lesssim \|\nabla v_{m_0}^j\|_{V_{2^*}(I^j)}\||\nabla|e^{it\Delta}w_n^k\|_{L^{\frac{(d+2)(q-1)}{q+1}}(E_{m_0}^j)} \qquad (8.162)\\
&\lesssim \{A(I)+1\}\||\nabla|e^{it\Delta}w_n^k\|_{L^2(E_{m_0}^j)}^{1-s_q}\||\nabla|e^{it\Delta}w_n^k\|_{V_{2+\frac{4}{d}}(\mathbb{R})}^{s_q} \\
&\lesssim \{A(I)+1\}(T_{m_0}^j)^{\frac{2(1-s_q)}{3(d+2)}}(R_{m_0}^j)^{\frac{(3d+2)(1-s_q)}{6(d+2)}}\|e^{it\Delta}w_n^k\|_{W_{2^*}(\mathbb{R})}^{\frac{1-s_q}{3}}\|\nabla w_n^k\|_{L^2}^{\frac{2(1-s_q)}{3}} \\
&\lesssim \{A(I)+1\}(T_{m_0}^j)^{\frac{2(1-s_q)}{3(d+2)}}(R_{m_0}^j)^{\frac{(3d+2)(1-s_q)}{6(d+2)}}\|e^{it\Delta}w_n^k\|_{W_{2^*}(\mathbb{R})}^{\frac{1-s_q}{3}} \le \delta.
\end{aligned}$$

Next, we assume that $\lambda_\infty^j = 0$. We see from Lemma 8.0.8 and the support property of $v_{m_0}^j$ that

$$\|\lambda_n^j\sigma_n^j v_{m_0}^j(t)\|_{L^2}^2 \lesssim \|\langle\nabla\rangle v_{m_0}^j(t)\|_{L^2}^2 \in L^1(\mathbb{R}). \tag{8.163}$$

We also see from Lemma 8.0.7 that for any $t \in \mathbb{R}$,

$$\lim_{n\to\infty}\|\lambda_n^j\sigma_n^j v_{m_0}^j(t)\|_{L^2} = 0. \tag{8.164}$$

Hence, Lebesgue's convergence theorem shows that

$$\lim_{n\to\infty}\|\lambda_n^j\sigma_n^j v_{m_0}^j\|_{L_{t,x}^2(I^j)}^2 = \lim_{n\to\infty}\int_{\mathbb{R}}\|\lambda_n^j\sigma_n^j v_{m_0}^j(t)\|_{L^2}^2\,dt = 0. \tag{8.165}$$

Furthermore, it follows from Hölder's inequality, (8.54), Sobolev's embedding, Lemma 8.0.8, (8.158), $\lambda_n^j \le 1$ and (8.165) that for any $k \ge 1$, there exists $N_3(\delta, I, j, k) \ge 1$ such that for any $n \ge N_3(\delta, I, j, k)$,

$$\begin{aligned}
&(\lambda_n^j)^{2-s_q}\|\sigma_n^j v_{m_0}^j (G_n^j)^{-1}|\nabla|e^{it\Delta}w_n^k\|_{L_{t,x}^{\frac{2(d+2)(q-1)}{d(q-1)+4}}(I^j)} \\
&\le (\lambda_n^j)^{1-s_q}\|\sigma_n^j v_{m_0}^j\|_{V_{2+\frac{4}{d}}(I^j)}^{1-s_q}\|\sigma_n^j v_{m_0}^j\|_{W_{2^*}(I^j)}^{s_q}\||\nabla|e^{it\Delta}w_n^k\|_{V_{2+\frac{4}{d}}(\mathbb{R})} \\
&\lesssim \|\lambda_n^j\sigma_n^j v_{m_0}^j\|_{V_{2+\frac{4}{d}}(I^j)}^{1-s_q}\{A(I)+1\}^{s_q} \qquad (8.166)\\
&\lesssim \|\lambda_n^j\sigma_n^j v_{m_0}^j\|_{L_t^2L_x^2(I^j)}^{\frac{1-s_q}{2}}(\lambda_n^j)^{\frac{1-s_q}{2}}\|\sigma_n^j v_{m_0}^j\|_{W_{2^*}(I^j)}^{\frac{1-s_q}{2}}\{A(I)+1\}^{s_q} \\
&\lesssim \|\lambda_n^j\sigma_n^j v_{m_0}^j\|_{L_t^2L_x^2(I^j)}^{\frac{1-s_q}{2}}\|\nabla\sigma_n^j v_{m_0}^j\|_{V_{2^*}(I^j)}^{\frac{1-s_q}{2}}\{A(I)+1\}^{s_q} \\
&\lesssim \|\lambda_n^j\sigma_n^j v_{m_0}^j\|_{L_t^2L_x^2(I_j)}^{\frac{1-s_q}{2}}\{A(I)+1\}^{\frac{1+s_q}{2}} \le \delta.
\end{aligned}$$

Case 2.2. Assume $s = 1$ and $\lambda_\infty^j = \infty$ in (8.153). We see from Hölder's inequality, (8.54), Sobolev's embedding, Strichartz' estimate and Lemma 8.0.7 that for any

$k \ge 1$, there exists $N_4(\delta, j, k) \ge 1$ such that for any $n \ge N_4(\delta, j, k)$,

$$(8.167)\quad \begin{aligned} &(\lambda_n^j)^{2-s_q} \|\sigma_n^j \widetilde{\psi}^j (G_n^j)^{-1} |\nabla| e^{it\Delta} w_n^k\|_{L_{t,x}^{\frac{2(d+2)(q-1)}{d(q-1)+4}}(I^j)} \\ &\le (\lambda_n^j)^{1-s_q} \|\sigma_n^j \widetilde{\psi}^j\|_{V_{2+\frac{4}{d}}(I^j)}^{1-s_q} \|\sigma_n^j \widetilde{\psi}^j\|_{W_{2^*}(I^j)}^{s_q} \| |\nabla| e^{it\Delta} w_n^k\|_{V_{2+\frac{4}{d}}(\mathbb{R})} \\ &\lesssim (\lambda_n^j)^{1-s_q} \|e^{it\Delta} \sigma_n^j \widetilde{u}^j\|_{V_{2+\frac{4}{d}}(I^j)}^{1-s_q} \|e^{it\Delta} \nabla \sigma_n^j \widetilde{u}^j\|_{V_{2^*}(I^j)}^{s_q} \\ &\lesssim \|\lambda_n^j \sigma_n^j \widetilde{u}^j\|_{L^2}^{1-s_q} \|\nabla \sigma_n^j \widetilde{u}^j\|_{L^2}^{s_q} \le \delta. \end{aligned}$$

Thus, putting the estimates (8.153), (8.154), (8.160), (8.161), (8.162), (8.166) and (8.167) together, we obtain the desired result (8.151). □

Now, for a number $k \ge 1$, we define an approximate solution ψ_n^{k-app} of ψ_n by

$$(8.168)\quad \psi_n^{k-app}(t) := \sum_{j=1}^{k} \psi_n^j(t) + e^{it\Delta} w_n^k.$$

Assume (8.127). Then, we have $I_{\max}^j = \mathbb{R}$ for all $j \ge 2$, which together with (8.82) shows that $I_{\max,n}^j = \mathbb{R}$ for all $j \ge 2$. Hence, for any numbers k and n, the maximal existence-interval of ψ_n^{k-app} is $I_{\max,n}^1 := \big((\lambda_n^1)^2 T_{\min}^1 + t_n^1, (\lambda_n^1)^2 T_{\max}^1 + t_n^1\big)$. Furthermore, we see from (8.75) that $0 \in I_{\max,n}^1$ for any sufficiently large number n.

LEMMA 8.0.21. *Assume (8.127), and let q denote one of the numbers $1 + \frac{4}{d}$, p and $2^* - 1$. Then, for any $j_0 \ge 2$, any $k > j_0$ and any $\delta > 0$, there exists a number $N(j_0, k, \delta)$ such that for any $n \ge N(j_0, k, \delta)$,*

$$(8.169)\quad \|\sum_{j=j_0}^{k} \psi_n^j\|_{W_{q+1}(\mathbb{R})}^{\frac{(d+2)(q-1)}{2}} \le \sum_{j=j_0}^{k} \|\psi_n^j\|_{W_{q+1}(\mathbb{R})}^{\frac{(d+2)(q-1)}{2}} + \delta,$$

$$(8.170)\quad \|\nabla \sum_{j=j_0}^{k} \psi_n^j\|_{V_{2+\frac{4}{d}}(\mathbb{R})}^{\frac{2(d+2)}{d}} \le \sum_{j=j_0}^{k} \|\nabla \psi_n^j\|_{V_{2+\frac{4}{d}}(\mathbb{R})}^{\frac{2(d+2)}{d}} + \delta.$$

PROOF OF LEMMA 8.0.21. Let j_0 and k be numbers with $2 \le j_0 < k$. Then, we can verify that there exists a constant $C(j_0, k)$ such that for any number n,

$$(8.171)\quad \begin{aligned} &\|\sum_{j=j_0}^{k} \psi_n^j\|_{W_{q+1}(\mathbb{R})}^{\frac{(d+2)(q-1)}{2}} \\ &\le \sum_{j=j_0}^{k} \|\psi_n^j\|_{W_{q+1}(\mathbb{R})}^{\frac{(d+2)(q-1)}{2}} + C(j_0, k) \sum_{j=2}^{k} \sum_{\substack{2 \le j' \le k; \\ j' \ne j}} \int_{\mathbb{R}} \int_{\mathbb{R}^d} |\psi_n^j|^{\frac{(d+2)(q-1)}{2} - 1} |\psi_n^{j'}| \, dx dt. \end{aligned}$$

Furthermore, the orthogonality (8.41) shows that for any $\delta > 0$, we can take a number $N(j_0, k, \delta)$ such that for any distinct numbers $j, j' \in \{2, \dots, k\}$ and any $n \ge N(j_0, k, \delta)$,

$$(8.172)\quad \int_{\mathbb{R}} \int_{\mathbb{R}^d} |\psi_n^j(x,t)|^{\frac{(d+2)(q-1)}{2} - 1} |\psi_n^{j'}(x,t)| \, dx dt \le \frac{\delta}{C(j_0, k) k^2}.$$

Putting (8.171) and (8.172) together, we obtain the desired result (8.169). Similarly, we can verify that (8.170) holds. □

LEMMA 8.0.22. *Assume* (8.127). *Then, for any interval I satisfying* (8.138), *we can take a constant $B(I) > 0$ with the following property: for any number k, there exists a number $N(k)$ such that for any $n \ge N(k)$,*

$$\|\psi_n^{k-app}\|_{W_{p+1}(I_n)\cap W_{2^*}(I_n)} + \|\langle\nabla\rangle\psi_n^{k-app}\|_{V_{2+\frac{4}{d}}(I_n)} \le B(I), \tag{8.173}$$

where $I_n := ((\lambda_n^1)^2 \inf I + t_n^1, (\lambda_n^1)^2 \sup I + t_n^1)$.

PROOF OF LEMMA 8.0.22. We consider the first term on the right-hand side of (8.173). Let q denote p or $2^* - 1$, and let J_0 be the number found in Lemma 8.0.14. Then, it follows from Lemma 8.0.21 with $\delta = 1$, Lemma 8.0.13, (8.54), Lemma 8.0.19, (8.136) and Lemma 8.0.14 that for any number k, there exists a number $N(k)$ such that for any $n \ge N(k)$,

$$\begin{aligned}
&\|\psi_n^{k-app}\|_{W_{q+1}(I_n)}^{\frac{(d+2)(q-1)}{2}} \\
&\lesssim \|\psi_n^1\|_{W_{q+1}(I_n)}^{\frac{(d+2)(q-1)}{2}} + \sum_{j=2}^{k} \|\psi_n^j\|_{W_{q+1}(\mathbb{R})}^{\frac{(d+2)(q-1)}{2}} + 1 + \|e^{it\Delta}w_n^k\|_{W_{q+1}(\mathbb{R})}^{\frac{(d+2)(q-1)}{2}} \\
&\lesssim \|\langle\nabla\rangle\widetilde{\psi}^1\|_{St(I)}^{\frac{(d+2)(q-1)}{2}} + \sum_{j=2}^{J_0} \|\langle\nabla\rangle\widetilde{\psi}^j\|_{St(\mathbb{R})}^{\frac{(d+2)(q-1)}{2}} + \sum_{j=J_0}^{\infty} \|\langle\nabla\rangle\widetilde{\psi}^j\|_{St(\mathbb{R})}^{\frac{(d+2)(q-1)}{2}} + 1 \\
&\lesssim A(I)^{\frac{(d+2)(q-1)}{2}} + J_0 + 1.
\end{aligned} \tag{8.174}$$

Thus, we have obtained the desired estimate for the first term. We can deal with the second term in a similar way. □

LEMMA 8.0.23. *Assume* (8.127), *and let q denote p or $2^* - 1$. Then, for any $\delta > 0$ and any interval I satisfying* (8.138), *we can take a number $K(\delta, I)$ with the following property: for any $k \ge K(\delta, I)$, there exists a number $N(\delta, I, k)$ such that for any $n \ge N(\delta, I, k)$ and $s \in \{0, 1\}$,*

$$\big\||\psi_n^{k-app}|^{q-1}|\nabla|^s e^{it\Delta}w_n^k\big\|_{L_{t,x}^{\frac{2(d+2)}{d+4}}(I_n)} \le \delta, \tag{8.175}$$

where $I_n := ((\lambda_n^1)^2 \inf I + t_n^1,\ (\lambda_n^1)^2 \sup I + t_n^1)$.

PROOF OF LEMMA 8.0.23. Let us begin with some preparation. We see from Sobolev's embedding, Lemma 8.0.13 and Lemma 8.0.14 that for any $\delta > 0$, there exists a number $K_1(\delta)$ such that

$$\sum_{j=K_1(\delta)}^{\infty} \|\psi_n^j\|_{W_{q+1}(\mathbb{R})}^{\frac{(d+2)(q-1)}{2}} \lesssim \sum_{j=K_1(\delta)}^{\infty} \|\langle\nabla\rangle\widetilde{\psi}^j\|_{St(\mathbb{R})}^{\frac{(d+2)(q-1)}{2}} \le \delta. \tag{8.176}$$

This together with Lemma 8.0.21 shows that for any $\delta > 0$ and any $k > K_1(\delta)$, there exists a number $N_1(\delta, k)$ such that for any $n \ge N_1(\delta, k)$,

$$\Big\| \sum_{j=K_1(\delta)+1}^{k} \psi_n^j\Big\|_{W_{q+1}(\mathbb{R})}^{\frac{(d+2)(q-1)}{2}} \lesssim \delta. \tag{8.177}$$

Furthermore, we see from Hölder's inequality, (8.54) and (8.55) that for any $\delta > 0$, there exists a number $K_2(\delta)$ with the following property: for any $k \geq K_2(\delta)$, there exists a number $N_2(\delta, k)$ such that for any $n \geq N_2(\delta, k)$ and $s \in \{0, 1\}$,

(8.178)
$$\begin{aligned}\big\|e^{it\Delta}w_n^k|\nabla|^s e^{it\Delta}w_n^k\big\|_{L_{t,x}^{\frac{2(d+2)(q-1)}{d(q-1)+4}}(\mathbb{R})}^{\frac{(d+2)(q-1)}{2}} &\leq \|e^{it\Delta}w_n^k\|_{W_{q+1}(\mathbb{R})}^{\frac{(d+2)(q-1)}{2}} \||\nabla|^s e^{it\Delta}w_n^k\|_{V_{2+\frac{4}{d}}(\mathbb{R})}^{\frac{(d+2)(q-1)}{2}} \\ &\lesssim \|e^{it\Delta}w_n^k\|_{W_{q+1}(\mathbb{R})}^{\frac{(d+2)(q-1)}{2}} \leq \delta.\end{aligned}$$

Now, we shall prove (8.175). First, we consider the case where $q \leq 2$. Then, we see from Hölder's inequality, (8.54), (8.177) and (8.178) that for any $\delta > 0$, any $k > \max\{K_1(\delta), K_2(\delta)\}$, any $n \geq \max\{N_1(\delta, k), N_2(\delta, k)\}$ and $s \in \{0, 1\}$,

$$\begin{aligned}&\||\psi_n^{k-app}|^{q-1}|\nabla|^s e^{it\Delta}w_n^k\|_{L_{t,x}^{\frac{2(d+2)}{d+4}}(I_n)}^{\frac{d+2}{2}} \\ &\leq \|\psi_n^{k-app}|\nabla|^s e^{it\Delta}w_n^k\|_{L_{t,x}^{\frac{2(d+2)(q-1)}{d(q-1)+4}}(I_n)}^{\frac{(d+2)(q-1)}{2}} \||\nabla|^s e^{it\Delta}w_n^k\|_{V_{2+\frac{4}{d}}(I_n)}^{\frac{(d+2)(2-q)}{2}} \\ &\lesssim \|\psi_n^{k-app}|\nabla|^s e^{it\Delta}w_n^k\|_{L_{t,x}^{\frac{2(d+2)(q-1)}{d(q-1)+4}}(I_n)}^{\frac{(d+2)(q-1)}{2}} \\ &\lesssim C(\delta)\sum_{j=1}^{K_1(\delta)} \|\psi_n^j|\nabla|^s e^{it\Delta}w_n^k\|_{L_{t,x}^{\frac{2(d+2)(q-1)}{d(q-1)+4}}(I_n)}^{\frac{(d+2)(q-1)}{2}} \qquad (8.179) \\ &\quad + \|\sum_{j=K_1(\delta)+1}^{k} \psi_n^j\|_{W_{q+1}(\mathbb{R})}^{\frac{(d+2)(q-1)}{2}} \||\nabla|^s e^{it\Delta}w_n^k\|_{V_{2+\frac{4}{d}}(\mathbb{R})}^{\frac{(d+2)(q-1)}{2}} \\ &\quad + \big\|e^{it\Delta}w_n^k|\nabla|^s e^{it\Delta}w_n^k\big\|_{L_{t,x}^{\frac{2(d+2)(q-1)}{d(q-1)+4}}(\mathbb{R})}^{\frac{(d+2)(q-1)}{2}} \\ &\lesssim C(\delta)K_1(\delta) \sup_{1\leq j\leq K_1(\delta)} \|\psi_n^j|\nabla|^s e^{it\Delta}w_n^k\|_{L_{t,x}^{\frac{2(d+2)(q-1)}{d(q-1)+4}}(I_n)}^{\frac{(d+2)(q-1)}{2}} + \delta,\end{aligned}$$

where $C(\delta)$ is some constant depending only on d, q and $K_1(\delta)$. Moreover, Lemma 8.0.20 shows that we can take a number $K_3(\delta, I)$ with the following property: for any $k \geq K_3(\delta, I)$, there exists a number $N_3(\delta, I, k)$ such that for any $1 \leq j \leq K_1(\delta)$, any $n \geq N_3(\delta, I, k)$ and $s \in \{0, 1\}$,

$$\big\|\psi_n^j|\nabla|^s e^{it\Delta}w_n^k\big\|_{L_{t,x}^{\frac{2(d+2)(q-1)}{d(q-1)+4}}(I_n)}^{\frac{(d+2)(q-1)}{2}} \leq \frac{\delta}{C(\delta)K_1(\delta)}. \qquad (8.180)$$

Putting (8.179) and (8.180) together, we obtain the desired result (8.175) in the case $q \leq 2$.

Next, we consider the case where $q > 2$. In this case, we see from Hölder's inequality and Lemma 8.0.22 that there exists $B(I) > 0$ with the following property: for any number k, there exists a number $N_4(k)$ such that for any $n \geq N_4(k)$ and

$s \in \{0,1\}$,

$$
\begin{aligned}
&\big\| |\psi_n^{k-app}|^{q-1} |\nabla|^s e^{it\Delta} w_n^k \big\|_{L_{t,x}^{\frac{2(d+2)}{d+4}}(I_n)} \\
&\le \| \psi_n^{k-app} |\nabla|^s e^{it\Delta} w_n^k \|_{L_{t,x}^{\frac{2(d+2)(q-1)}{d(q-1)+4}}(I_n)} \| \psi_n^{k-app} \|_{W_{q+1}}^{q-2} \qquad (8.181)\\
&\le \| \psi_n^{k-app} |\nabla|^s e^{it\Delta} w_n^k \|_{L_{t,x}^{\frac{2(d+2)(q-1)}{d(q-1)+4}}(I_n)} B(I)^{q-2}.
\end{aligned}
$$

Then, we can obtain the desired estimate (8.175) in a way similar to the case $q \le 2$. □

The following lemma enables us to control the error term of the approximate solution:

LEMMA 8.0.24. *Assume* (8.127). *Then, for any $\delta \in (0,1)$ and any interval I satisfying* (8.138), *there exist numbers k_0 (depending on δ and I) and $N(\delta, I)$ such that for any $n \ge N(\delta, I)$,*

$$
\| \langle \nabla \rangle e[\psi_n^{k_0-app}] \|_{L_{t,x}^{\frac{2(d+2)}{d+4}}(I_n)} \le \delta, \qquad (8.182)
$$

where $I_n := ((\lambda_n^1)^2 \inf I + t_n^1, (\lambda_n^1)^2 \sup I + t_n^1)$.

PROOF OF LEMMA 8.0.24. Let q denote p or $2^* - 1$. Then, we see from (8.55), Lemma 8.0.22 and Lemma 8.0.23 that for any $\delta \in (0,1)$ and any interval I satisfying (8.138), there exist a number $k_0 \ge 1$ (depending on δ and I), a constant $B(I) > 0$ (independent of δ) and a number $N_1(\delta, I) \ge 1$ such that if $n \ge N_1(\delta, I)$, then

(8.183)

$$
\| e^{it\Delta} w_n^{k_0} \|_{W_{q+1}(\mathbb{R})} \le \frac{\delta}{(1+B(I))^{2^*}},
$$

(8.184)

$$
\| \psi_n^{k_0-app} \|_{W_{q+1}(I_n)} + \| \langle \nabla \rangle \psi_n^{k_0-app} \|_{V_{2+\frac{4}{d}}(I_n)} \le B(I),
$$

(8.185)

$$
\big\| |\psi_n^{k_0-app}|^{q-1} e^{it\Delta} w_n^{k_0} \big\|_{L_{t,x}^{\frac{2(d+2)}{d+4}}(I_n)} + \big\| |\psi_n^{k_0-app}|^{q-1} |\nabla| e^{it\Delta} w_n^{k_0} \big\|_{L_{t,x}^{\frac{2(d+2)}{d+4}}(I_n)} \le \delta.
$$

We rewrite the "error term" $e[\psi_n^{k_0-app}]$ as follows:

$$
\begin{aligned}
e[\psi_n^{k_0-app}] &= \sum_{j=1}^{k_0} \Big(i\frac{\partial \psi_n^j}{\partial t} + \Delta \psi_n^j \Big) + F\big[\sum_{j=1}^{k_0} \psi_n^j\big] \qquad (8.186)\\
&\quad + F\big[\psi_n^{k_0-app}\big] - F\big[\psi_n^{k_0-app} - e^{it\Delta} w_n^{k_0}\big].
\end{aligned}
$$

Then, we have

$$\begin{aligned}
&\|\langle\nabla\rangle e[\psi_n^{k_0-app}]\|_{L_{t,x}^{\frac{2(d+2)}{d+4}}(I_n)} \\
&\le \|\langle\nabla\rangle \Big\{ \sum_{j=1}^{k_0}\Big(i\frac{\partial\psi_n^j}{\partial t}+\Delta\psi_n^j\Big)+F\big[\sum_{j=1}^{k_0}\psi_n^j\big] \Big\} \|_{L_{t,x}^{\frac{2(d+2)}{d+4}}(I_n)} \\
&\quad + \big\|\langle\nabla\rangle \Big\{ F\big[\psi_n^{k_0-app}\big]-F\big[\psi_n^{k_0-app}-e^{it\Delta}w_n^{k_0}\big] \Big\} \big\|_{L_{t,x}^{\frac{2(d+2)}{d+4}}(I_n)}.
\end{aligned} \tag{8.187}$$

We consider the first term on the right-hand side of (8.187). It follows from (8.81), Lemma 8.0.11 and Lemma 8.0.12 that there exists a number $N_2(\delta,I)$ such that for any $n\ge N_2(\delta,I)$,

$$\begin{aligned}
&\|\langle\nabla\rangle \Big\{ \sum_{j=1}^{k_0}\Big(i\frac{\partial\psi_n^j}{\partial t}+\Delta\psi_n^j\Big)+F\big[\sum_{j=1}^{k_0}\psi_n^j\big] \Big\} \|_{L_{t,x}^{\frac{2(d+2)}{d+4}}(I_n)} \\
&\le \|\langle\nabla\rangle\Big\{\sum_{j=1}^{k_0}F[\psi_n^j]-F\big[\sum_{j=1}^{k_0}\psi_n^j\big]\Big\}\|_{L_{t,x}^{\frac{2(d+2)}{d+4}}(I_n)}+\delta.
\end{aligned} \tag{8.188}$$

Then, we see from elementary computations that for $q=p$ or $q=2^*-1$,

$$\begin{aligned}
&\|\sum_{j=1}^{k_0}|\psi_n^j|^{q-1}\psi_n^j-|\sum_{j=1}^{k_0}\psi_n^j|^{q-1}\sum_{j=1}^{k_0}\psi_n^j\|_{L_{t,x}^{\frac{2(d+2)}{d+4}}(I_n)} \\
&\lesssim \sum_{j=1}^{k_0}\sum_{\substack{1\le k\le k_0\\ k\ne j}} \||\psi_n^j|^{q-1}\psi_n^k\|_{L_{t,x}^{\frac{2(d+2)}{d+4}}(I_n)}
\end{aligned} \tag{8.189}$$

and

$$\begin{aligned}
&\|\nabla\Big\{\sum_{j=1}^{k_0}|\psi_n^j|^{q-1}\psi_n^j-|\sum_{j=1}^{k_0}\psi_n^j|^{q-1}\sum_{j=1}^{k_0}\psi_n^j\Big\}\|_{L_{t,x}^{\frac{2(d+2)}{d+4}}(I_n)} \\
&\lesssim \sum_{j=1}^{k_0}\sum_{\substack{1\le k\le k_0\\ k\ne j}} \||\psi_n^j|^{q-1}\nabla\psi_n^k\|_{L_{t,x}^{\frac{2(d+2)}{d+4}}(I_n)}.
\end{aligned} \tag{8.190}$$

Here, when $q>2$, we must add the following term to the right-hand side of (8.190):

$$\sum_{i=1}^{k_0}\sum_{j=1}^{k_0}\sum_{\substack{1\le k\le k_0\\ k\ne j}} \||\psi_n^i|^{q-2}|\nabla\psi_n^j||\psi_n^k|\|_{L_{t,x}^{\frac{2(d+2)}{d+4}}(I_n)}. \tag{8.191}$$

We see from the orthogonality (8.41) that there exists a number $N_3(\delta,I)$ such that for any distinct numbers $j,k\in\{1,\dots,k_0\}$ and any $n\ge N_3(\delta,I)$,

$$\||\psi_n^j|^{q-1}\psi_n^k\|_{L_{t,x}^{\frac{2(d+2)}{d+4}}(I_n)}+\||\psi_n^j|^{q-1}\nabla\psi_n^k\|_{L_{t,x}^{\frac{2(d+2)}{d+4}}(I_n)}\le\frac{\delta}{k_0^2}. \tag{8.192}$$

When $q > 2$, we also see that for any number $i \in \{1, \dots, k_0\}$, any distinct numbers $j, k \in \{1, \dots, k_0\}$ and any $n \ge N_3(\delta, I)$,

$$\| |\psi_n^i|^{q-2} \nabla \psi_n^j \, \psi_n^k \|_{L_{t,x}^{\frac{2(d+2)}{d+4}}(I_n)} \le \frac{\delta}{k_0^3}. \tag{8.193}$$

Putting the estimates (8.188) through (8.193) together, we find that for any $n \ge \max\{N_2(\delta, I), N_3(\delta, I)\}$,

$$\|\langle \nabla \rangle \Big\{ \sum_{j=1}^{k_0} \Big(i \frac{\partial \psi_n^j}{\partial t} + \Delta \psi_n^j \Big) + F\big[\sum_{j=1}^{k_0} \psi_n^j\big] \Big\} \|_{L_{t,x}^{\frac{2(d+2)}{d+4}}(I_n)} \lesssim \delta. \tag{8.194}$$

We move on to the second term on the right-hand side of (8.187). We can verify that for $q = p$ or $q = 2^* - 1$, and any number $n \ge 1$,

$$\begin{aligned}
&\|\langle \nabla \rangle \Big\{ |\psi_n^{k_0-app}|^{q-1} \psi_n^{k_0-app} - |\psi_n^{k_0-app} - e^{it\Delta} w_n^{k_0}|^{q-1} (\psi_n^{k_0-app} - e^{it\Delta} w_n^{k_0}) \Big\} \|_{L_{t,x}^{\frac{2(d+2)}{d+4}}(I_n)} \\
&\lesssim \| |\psi_n^{k_0-app}|^{q-1} e^{it\Delta} w_n^{k_0} \|_{L_{t,x}^{\frac{2(d+2)}{d+4}}(I_n)} + \| |e^{it\Delta} w_n^{k_0}|^{q-1} \psi_n^{k_0-app} \|_{L_{t,x}^{\frac{2(d+2)}{d+4}}(I_n)} \\
&\quad + \| |\psi_n^{k_0-app}|^{q-1} \nabla e^{it\Delta} w_n^{k_0} \|_{L_{t,x}^{\frac{2(d+2)}{d+4}}(I_n)} + \| |e^{it\Delta} w_n^{k_0}|^{q-1} \nabla \psi_n^{k_0-app} \|_{L_{t,x}^{\frac{2(d+2)}{d+4}}(I_n)} \\
&\quad + \|\langle \nabla \rangle \{ |e^{it\Delta} w_n^{k_0}|^{q-1} e^{it\Delta} w_n^{k_0} \} \|_{L_{t,x}^{\frac{2(d+2)}{d+4}}(I_n)}.
\end{aligned} \tag{8.195}$$

When $q > 2$, we must add the following terms to the right-hand side of (8.195):

$$\| |\psi_n^{k_0-app}|^{q-2} e^{it\Delta} w_n^{k_0} \nabla \psi_n^{k_0-app} \|_{L_{t,x}^{\frac{2(d+2)}{d+4}}(I_n)}, \tag{8.196}$$

$$\| |e^{it\Delta} w_n^{k_0}|^{q-2} \psi_n^{k_0-app} \nabla e^{it\Delta} w_n^{k_0} \|_{L_{t,x}^{\frac{2(d+2)}{d+4}}(I_n)}. \tag{8.197}$$

Furthermore, using (8.185), Hölder's inequality, (8.183), (8.184) and (8.54), we can verify that the right-hand side of (8.195) vanishes as n tends to the infinity (cf. the proof of (3.10) in [**21**]). Thus, we conclude that

$$\lim_{n \to \infty} \big\| \langle \nabla \rangle \{ F[\psi_n^{k_0-app}] - F[\psi_n^{k_0-app} - e^{it\Delta} w_n^{k_0}] \} \big\|_{L_{t,x}^{\frac{2(d+2)}{d+4}}(I_n)} = 0. \tag{8.198}$$

Putting (8.187), (8.194) and (8.198) together, we find that the desired estimate (8.182) holds. □

PROPOSITION 8.0.25. *Assume* (8.127). *Then,* $\lambda_\infty^1 \in \{0, 1\}$, *and for any* $T \in I_{\max}^1 = (T_{\min}^1, T_{\max}^1)$,

$$\|\sigma_\infty^1 \widetilde{\psi}^1\|_{W^1([T, T_{\max}^1))} = \infty. \tag{8.199}$$

Furthermore, the following holds:

$$\begin{cases} \mathcal{S}_\omega(\sigma_\infty^1 \widetilde{\psi}^1) \ge m_\omega & \text{if } \lambda_\infty^1 = 1, \\ \mathcal{H}^\ddagger(\sigma_\infty^1 \widetilde{\psi}^1) \ge \frac{1}{d} \sigma^{\frac{d}{2}} > m_\omega & \text{if } \lambda_\infty^1 = 0. \end{cases} \tag{8.200}$$

PROOF OF PROPOSITION 8.0.25. First, we assume $\lambda_\infty^1 \in \{0,1\}$ and prove (8.199). Suppose for contradiction that (8.199) failed for some $T \in I_{\max}^1$. Then, we see from the well-posedness theory that $T_{\max}^1 = \infty$ and for any $\tau \in (T_{\min}^1, \infty)$,

$$\|\sigma_\infty^1 \widetilde{\psi}^1\|_{\mathcal{W}^1([\tau,\infty))} < \infty. \tag{8.201}$$

In order to derive a contradiction, we consider the approximate solution ψ_n^{k-app} defined by (8.168). Note that for any numbers k and n, the maximal existence interval of ψ_n^{k-app} is identical to $I_{\max,n}^1 := ((\lambda_n^1)^2 T_{\min}^1 + t_n^1, \infty)$.

We shall show that there exists a constant $B > 0$ with the following property: for any number k, there exists a number $N(k)$ such that for any $n \ge N(k)$,

$$\|\psi_n^{k-app}\|_{V_{2+\frac{4}{d}}([0,\infty)) \cap W_{2^*}([0,\infty))} \le B. \tag{8.202}$$

If $-\frac{t_\infty^1}{(\lambda_\infty^1)^2} = -\infty$, then we see from (8.201) and (8.78) that Lemma 8.0.22 is available on the whole interval $\mathbb{R}$. Thus, we can take a constant $B_1 > 0$ with the following property: for any number k, there exists a number $N_1(k)$ such that for any $n \ge N_1(k)$,

$$\|\psi_n^{k-app}\|_{V_{2+\frac{4}{d}}(\mathbb{R}) \cap W_{2^*}(\mathbb{R})} \le B_1. \tag{8.203}$$

On the other hand, if $-\frac{t_\infty^1}{(\lambda_\infty^1)^2} \neq -\infty$, then it follows from the construction of $\widetilde{\psi}^1$ (see (8.72)) that $-\frac{t_\infty^1}{(\lambda_\infty^1)^2} \in (T_{\min}^1, \infty) \cup \{\infty\}$. This implies that there exist $\tau_0 \in (T_{\min}^1, \infty)$ and a number N_0 such that $\tau_0 < -\frac{t_n^1}{(\lambda_n^1)^2}$ for any $n \ge N_0$, so that $[0,\infty) \subset ((\lambda_n^1)^2 \tau_0 + t_n^1, \infty) \subset I_{\max,n}^1$. Furthermore, we see from Lemma 8.0.22 and (8.201) with $\tau = \tau_0$ that there exists a constant $B_2 > 0$ with the following property: for any number k, there exists a number $N_2(k)$ such that for any $n \ge \max\{N_0, N_2(k)\}$,

$$\|\psi_n^{k-app}\|_{V_{2+\frac{4}{d}}([0,\infty)) \cap W_{2^*}([0,\infty))} \le B_2. \tag{8.204}$$

Thus, (8.203) and (8.204) give the desired result (8.202).

Now, we shall finish the proof of (8.199) for $\lambda_\infty^1 \in \{0,1\}$. Note that it follows from (8.43), (8.46) and Lemma 8.0.8 that

(8.205)

$$\begin{aligned}\|\psi_n(0) - \psi_n^{k-app}(0)\|_{H^1} &= \Big\| \sum_{j=1}^k g_n^j \sigma_n^j e^{-i\frac{t_n^j}{(\lambda_n^j)^2}\Delta} \widetilde{u}^j - \sum_{j=1}^k g_n^j \sigma_n^j \widetilde{\psi}^j\Big(-\frac{t_n^j}{(\lambda_n^j)^2}\Big)\Big\|_{H^1} \\ &\lesssim \sum_{j=1}^k \Big\| e^{-\frac{t_n^j}{(\lambda_n^j)^2}\Delta} \widetilde{u}^j - \widetilde{\psi}^j\Big(-\frac{t_n^j}{(\lambda_n^j)^2}\Big)\Big\|_{H^1}.\end{aligned}$$

This estimate together with (8.76) shows that for any number k, there exists a number $N_3(k)$ such that

$$\sup_{n \ge N_3(k)} \|\psi_n(0) - \psi_n^{k-app}(0)\|_{H^1} \le 1. \tag{8.206}$$

We also see from (8.38) that

$$\sup_{n \ge 1} \|\psi_n\|_{L_t^\infty H_x^1([0,\infty))} \lesssim 1. \tag{8.207}$$

Let $\delta_0 > 0$ be a constant determined by the long-time perturbation theory (Lemma E.0.1) together with (8.202), (8.206) and (8.207). Then, we see from Sobolev's embedding, Strichartz' estimate, (8.205) and (8.76) that for any number k, there exists a number $N_4(k)$ such that for any $n \ge N_4(k)$,

(8.208)

$$\begin{aligned}&\|e^{it\Delta}\{\psi_n(0) - \psi_n^{k-app}(0)\}\|_{V_{2+\frac{4}{d}}([0,\infty))\cap W_{2^*}([0,\infty))} \\ &\lesssim \|\langle\nabla\rangle e^{it\Delta}\{\psi_n(0) - \psi_n^{k-app}(0)\}\|_{St([0,\infty))} \lesssim \|\psi_n(0) - \psi_n^{k-app}(0)\|_{H^1} \ll \delta_0.\end{aligned}$$

Moreover, it follows from Lemma 8.0.24 that we can take numbers k_0 and N_5 such that for any $n \ge N_5$,

$$\|\langle\nabla\rangle e[\psi_n^{k_0-app}]\|_{L_{t,x}^{\frac{2(d+2)}{d+4}}([0,\infty))} \le \delta_0. \tag{8.209}$$

Thus, applying the long-time perturbation theory (Lemma E.0.1) to ψ_n and $\psi_n^{k_0-app}$ with $n \ge \max\{N(k_0), N_3(k_0), N_4(k_0), N_5\}$, we find that

$$\|\langle\nabla\rangle\psi_n\|_{St([0,\infty))} < \infty. \tag{8.210}$$

However, this contradicts (8.34). Thus, we have shown that (8.199) holds when $\lambda_\infty^1 \in \{0,1\}$.

Next, we assume that $\lambda_\infty^1 = \infty$. Then, (8.202) holds without the hypothesis (8.201). Furthermore, the same argument as (8.205) through (8.210) shows that the case $\lambda_\infty^1 = \infty$ never happens.

It remains to prove (8.200). Suppose to the contrary that the claim (8.200) was false. Then, it follows from (8.74), Theorem 1.0.2, Theorem 1.0.3, Lemma 8.0.18 and (8.199) that $\|\sigma_\infty^j\widetilde{\psi}^j\|_{St([0,\infty))} < \infty$. However, this contradicts (8.199). Thus, we have proved that the claim (8.200) is true. □

8.0.2.5. *Critical element and completion of the proof.* We shall show the existence of "critical element" (see Lemma 8.0.29 below). Furthermore, we will derive a contradiction by using it under the hypothesis (8.30), which completes the proof of Proposition 8.0.6.

Let us begin by defining the functionals $\mathcal{S}_\omega^j$ and $\mathcal{I}_\omega^j$. For each $j \ge 1$, we define

$$\mathcal{S}_\omega^j := \begin{cases} \mathcal{S}_\omega & \text{if} \quad \lambda_\infty^j \in \{1,\infty\}, \\ \mathcal{H}^\ddagger & \text{if} \quad \lambda_\infty^j = 0, \end{cases} \qquad \mathcal{I}_\omega^j := \begin{cases} \mathcal{I}_\omega & \text{if} \quad \lambda_\infty^j \in \{1,\infty\}, \\ \mathcal{I}^\ddagger & \text{if} \quad \lambda_\infty^j = 0. \end{cases} \tag{8.211}$$

LEMMA 8.0.26. *Assume* (8.127). *Then, for any $\delta > 0$ and any number $k \ge 2$, there exists a number $N(\delta,k)$ such that for any $t \in \mathbb{R}$ and any $n \ge N(\delta,k)$,*

$$\sum_{j=2}^k \mathcal{I}_\omega^j(\sigma_\infty^j\widetilde{\psi}^j(t)) + \mathcal{I}_\omega(w_n^k) \le \delta. \tag{8.212}$$

PROOF OF LEMMA 8.0.26. We see from (8.31), (8.33) and (8.52) that for any $\delta > 0$ and any number k, there exists a number $N_1(\delta,k)$ such that for any $n \ge N_1(\delta,k)$,

$$\mathcal{S}_\omega\big(g_n^1\sigma_n^1 e^{-i\frac{t_n^1}{(\lambda_n^1)^2}\Delta}\widetilde{u}^1\big) + \sum_{j=2}^k \mathcal{S}_\omega\big(g_n^j\sigma_n^j e^{-i\frac{t_n^j}{(\lambda_n^j)^2}\Delta}\widetilde{u}^j\big) + \mathcal{S}_\omega(w_n^k) \le m_\omega + \delta. \tag{8.213}$$

If $\lambda_{\infty}^{1}=\lim_{n\to\infty}\lambda_{n}^{1}=0$, then Lemma 8.0.7 shows that for each $t\in I_{\max}^{1}$,

$$\lim_{n\to\infty}\|g_{n}^{1}\sigma_{n}^{1}\widetilde{\psi}^{1}(t)\|_{L^{2}}=0. \tag{8.214}$$

We see from Lemma 8.0.18, (8.120) in Lemma 8.0.16, (8.79), (8.213), (8.214) and (8.200) in Proposition 8.0.25 that for any $\delta>0$ and any number k, there exists $N_{2}(\delta,k)$ such that for any $t\in\mathbb{R}$ and any $n\geq N_{2}(\delta,k)$,

$$\begin{aligned}\sum_{j=2}^{k}\mathcal{I}_{\omega}^{j}(\sigma_{\infty}^{j}\widetilde{\psi}^{j}(t))+\mathcal{I}_{\omega}(w_{n}^{k})&\leq\sum_{j=2}^{k}\mathcal{S}_{\omega}^{j}(\sigma_{\infty}^{j}\widetilde{\psi}^{j})+\mathcal{S}_{\omega}(w_{n}^{k})\\&\leq\sum_{j=2}^{k}\mathcal{S}_{\omega}^{j}(g_{n}^{j}\sigma_{n}^{j}e^{-i\frac{t_{n}^{j}}{(\lambda_{n}^{j})^{2}}\Delta}\widetilde{u}^{j})+\delta+\mathcal{S}_{\omega}(w_{n}^{k})\\&\leq m_{\omega}-\mathcal{S}_{\omega}^{1}(\sigma_{\infty}^{1}\widetilde{\psi}^{1})+2\delta\leq2\delta.\end{aligned} \tag{8.215}$$

Thus, we have completed the proof. □

Put $\tau_{n}^{1}:=-\frac{t_{n}^{j}}{(\lambda_{n}^{1})^{2}}$. Then, passing to a subsequence, we may assume that

$$\tau_{\infty}^{1}:=\lim_{n\to\infty}\tau_{n}^{1}\in\mathbb{R}\cup\{\pm\infty\}. \tag{8.216}$$

LEMMA 8.0.27. *Assume (8.127). Then, we have that* $\lambda_{\infty}^{1}=1$. *Furthermore, there exists a time* $T\in I_{\max}^{1}$ *such that*

$$\inf_{t\in[T,T_{\max}^{1})}\widetilde{d}_{\omega}(\widetilde{\psi}^{1}(t))\geq\frac{R_{*}}{2} \tag{8.217}$$

and

$$\inf_{t\in[T,T_{\max}^{1})}\mathcal{K}(\widetilde{\psi}^{1}(t))\geq\frac{\kappa_{1}(R_{*})}{2}, \tag{8.218}$$

where $\kappa_{1}(R_{*})$ *is the constant appearing in* (8.37).

PROOF OF LEMMA 8.0.27. Proposition 8.0.25 together with (8.77) shows that $\lambda_{\infty}^{1}\neq\infty$ and $\tau_{\infty}^{1}\neq\infty$. If $\tau_{\infty}^{1}=-\infty$, then it follows from (8.78) that $T_{\min}^{1}=-\infty$ and $\|\sigma_{\infty}^{1}\widetilde{\psi}^{1}\|_{W^{1}((-\infty,T])}<\infty$ for any $T<T_{\max}^{1}$. If $\tau_{\infty}^{1}\in\mathbb{R}$, then we see from the construction of $\widetilde{\psi}^{1}$ (see (8.72)) that $\tau_{\infty}^{1}\in I_{\max}^{1}$. Put

$$\tau_{\min}^{1}:=\begin{cases}\dfrac{T_{\min}^{1}+\tau_{\infty}^{1}}{2}&\text{if }T_{\min}^{1}>-\infty,\\\tau_{\infty}^{1}-1&\text{if }\tau_{\infty}^{1}\in\mathbb{R}\text{ and }T_{\min}^{1}=-\infty,\\-\infty&\text{if }\tau_{\infty}^{1}=-\infty.\end{cases} \tag{8.219}$$

Then, $T_{\min}^{1}\leq\tau_{\min}^{1}\leq\tau_{\infty}^{1}$, and we can take a number N_{1} such that for any $n\geq N_{1}$,

$$\tau_{\min}^{1}<-\frac{t_{n}^{1}}{(\lambda_{n}^{1})^{2}}<T_{\max}^{1}. \tag{8.220}$$

Note that $0\in I_{\max,n}^{1}$ for all $n\geq N_{1}$. Furthermore, for each $T\in(\tau_{\min}^{1},T_{\max}^{1})$,

$$\|\sigma_{\infty}^{1}\widetilde{\psi}^{1}\|_{W^{1}((\tau_{\min}^{1},T])}<\infty, \tag{8.221}$$

which together with Lemma 8.0.19 shows that for any $T \in (\tau_{\min}^1, T_{\max}^1)$, there exists a constant $A(T) > 0$ such that

$$\|\langle\nabla\rangle\widetilde{\psi}^1\|_{St((\tau_{\min}^1,T])} \le A(T). \tag{8.222}$$

Sobolev's embedding also gives us that

$$\|\sigma_\infty^1\widetilde{\psi}^1\|_{W^1((\tau_{\min}^1,T])} \lesssim \|\langle\nabla\rangle\widetilde{\psi}^1\|_{St((\tau_{\min}^1,T])} \le A(T). \tag{8.223}$$

Now, we consider the approximate solution ψ_n^{k-app} defined by (8.168). Lemma 8.0.22 together with (8.223) shows that for any $T \in (\tau_{\min}^1, T_{\max}^1)$, there exists $B(T) > 0$ with the following property: for any number k, there exists a number $N_2(k)$ such that for any $n \ge N_2(k)$,

$$\|\psi_n^{k-app}\|_{V_{2+\frac{4}{d}}(I_n(T))\cap W_{2^*}(I_n(T))} + \|\langle\nabla\rangle\psi_n^{k-app}\|_{L_t^\infty L_x^2(I_n(T))} \le B(T), \tag{8.224}$$

where

$$I_n(T) := \left((\lambda_n^1)^2\tau_{\min}^1 + t_n^1,\, (\lambda_n^1)^2 T + t_n^1\right]. \tag{8.225}$$

Lemma 8.0.24 together with (8.223) shows that for any $\delta \in (0,1)$ and any $T \in (\tau_{\min}^1, T_{\max}^1)$, there exists numbers k_0 (depending on δ and T) and $N_3(\delta, T)$ such that for any $n \ge N_3(\delta, T)$,

$$\|\langle\nabla\rangle e[\psi_n^{k_0-app}]\|_{L_{t,x}^{\frac{2(d+2)}{d+4}}(I_n(T))} \le \delta. \tag{8.226}$$

We also see from (8.43), Lemma 8.0.8 and (8.76) that for any number k and any $\gamma > 0$, there exists a number $N_4(k,\gamma)$ such that for any $n \ge N_4(k,\gamma)$,

(8.227)

$$\begin{aligned}\big\|\psi_n(0) - \psi_n^{k-app}(0)\big\|_{H^1} &= \Big\|\sum_{j=1}^k g_n^j\sigma_n^j e^{-i\frac{t_n^j}{(\lambda_n^j)^2}\Delta}\widetilde{u}^j - \sum_{j=1}^k g_n^j\sigma_n^j\widetilde{\psi}^j\Big(-\frac{t_n^j}{(\lambda_n^j)^2}\Big)\Big\|_{H^1} \\ &\lesssim \sum_{j=1}^k \Big\|e^{-i\frac{t_n^j}{(\lambda_n^j)^2}\Delta}\widetilde{u}^j - \widetilde{\psi}^j\Big(-\frac{t_n^j}{(\lambda_n^j)^2}\Big)\Big\|_{H^1} \le \gamma.\end{aligned}$$

Furthermore, it follows from Strichartz' estimate and (8.227) that for any number k and any $\delta > 0$, there exists a number $N_5(k,\delta)$ such that for any $n \ge N_5(k,\delta)$,

$$\big\|\langle\nabla\rangle e^{it\Delta}\{\psi_n(0) - \psi_n^{k-app}(0)\}\big\|_{V_{p+1}(\mathbb{R})} \lesssim \big\|\psi_n(0) - \psi_n^{k-app}(0)\big\|_{H^1} \le \delta. \tag{8.228}$$

The long-time perturbation theory (Lemma E.0.1) together with (8.224), (8.226), (8.227) and (8.228) shows that for any $T \in (\tau_{\min}^1, T_{\max}^1)$, there exist constants $\delta_0(T,\gamma) > 0$ and $C(T,\gamma) > 0$ with the following properties: for any $\delta \in (0, \delta_0(T,\gamma))$, we can take numbers k_0 (depending on δ and T) and $N_6(\delta, T)$ such that for any $n \ge N_6(\delta, T)$,

$$\big\|\langle\nabla\rangle\psi_n\big\|_{St(I_n(T))} \le C(T,\gamma), \tag{8.229}$$

$$\sup_{t\in I_n(T)} \|\langle\nabla\rangle\{\psi_n(t) - \psi_n^{k_0-app}(t)\}\|_{L^2} \le C(T,\gamma)\gamma. \tag{8.230}$$

Here, $C(T,\gamma)$ is non-decreasing with respect to γ. Furthermore, it follows from (8.230), Lemma 8.0.7 and Lemma 8.0.26 that for any $T \in (\tau_{\min}^1, T_{\max}^1)$, any $\delta \in$

$(0, \delta_0(T,\gamma))$ and the number k_0 determined by T and δ, we can take a number $N_7(\delta, T, \omega)$ such that for any $n \ge N_7(\delta, T, \omega)$ and $s \in \{0, 1\}$,

$$\begin{aligned}
&\| |\nabla|^s \{\psi_n((\lambda_n^1)^2T + t_n^1) - \psi_n^1((\lambda_n^1)^2T + t_n^1)\}\|_{L^2}^2 \\
&\lesssim \| |\nabla|^s \{\psi_n((\lambda_n^1)^2T + t_n^1) - \psi_n^{k_0-app}((\lambda_n^1)^2T + t_n^1)\}\|_{L^2}^2 \\
&\quad + \Big\| |\nabla|^s \sum_{j=2}^{k_0} \psi_n^j((\lambda_n^1)^2T + t_n^1)\Big\|_{L^2}^2 + \| |\nabla|^s w_n^{k_0}\|_{L^2}^2 \\
&\lesssim C(T,\gamma)^2\gamma^2 + k_0 \sum_{j=2}^{k_0} \Big\| |\nabla|^s \psi_n^j((\lambda_n^1)^2T + t_n^1)\Big\|_{L^2}^2 + \| |\nabla|^s w_n^{k_0}\|_{L^2}^2 + \delta \\
&= C(T,\gamma)^2\gamma^2 + k_0 \sum_{j=2}^{k_0} \Big\| |\nabla|^s g_n^j \sigma_n^j \widetilde{\psi}^j(T)\Big\|_{L^2}^2 + \| |\nabla|^s w_n^{k_0}\|_{L^2}^2 + \delta \\
&\lesssim C(T,\gamma)^2\gamma^2 + \delta.
\end{aligned} \tag{8.231}$$

We shall show that $\lambda_\infty^1 = 1$. Suppose for contradiction that $\lambda_\infty^1 = 0$. Fix a time $T \in (\tau_{\min}^1, T_{\max}^1)$. Then, it follows from Lemma 8.0.7 that

$$\lim_{n\to\infty} \big\|\psi_n^1((\lambda_n^1)^2T + t_n^1)\big\|_{L^2} = \lim_{n\to\infty} \big\|g_n^1\sigma_n^1\widetilde{\psi}^1(T)\big\|_{L^2} = 0. \tag{8.232}$$

Furthermore, we see from $\mathcal{M}(\psi_n) \equiv \mathcal{M}(\Phi_\omega)$, (8.231) and (8.232) that for any sufficiently small $\omega > 0$ and any $\delta \in (0, \delta_0(T,\gamma))$,

$$\|\Phi_\omega\|_{L^2}^2 = \lim_{n\to\infty} \|\psi_n((\lambda_n^1)^2T + t_n^1)\|_{L^2}^2 \lesssim C(T,\gamma)^2\gamma^2 + \delta. \tag{8.233}$$

Taking $\gamma \to 0$ and $\delta \to 0$, we deduce that $\Phi_\omega \equiv 0$. However, this is a contradiction. Thus, we find that $\lambda_\infty^1 = 1$, so that $\lambda_n^1 \equiv 1$, $\tau_\infty^1 = -t_\infty^1$ and $\sigma_n^1 \equiv 1$.

Now, we are in a position to prove (8.217) and (8.218). We first consider the case $\tau_\infty^1 = -\infty$. In this case, we have $t_\infty^1 = \infty$. Suppose for contradiction that (8.217) failed. Then, for any $T \in (\tau_{\min}^1, T_{\max}^1)$, we can take $T' \in [T, T_{\max}^1)$ such that

$$\widetilde{d}_\omega(\widetilde{\psi}^1(T')) \le \frac{2}{3}R_*. \tag{8.234}$$

Since $\lim_{n\to\infty} t_n^1 = \infty$, we can take a number $N_8(T')$ such that $T' + t_n^1 \ge 0$ for any $n \ge N_8(T')$. Thus, we find from (8.35), (8.231) with $\lambda_n^1 \equiv 1$ and (8.234) that

$$\begin{aligned}
R_* &\le \widetilde{d}_\omega(\psi_n(T' + t_n^1)) \le \widetilde{d}_\omega(\psi_n^1(T' + t_n^1)) + o_\gamma(1) + o_\delta(1) \\
&= \widetilde{d}_\omega(\widetilde{\psi}^1(T')) + o_\gamma(1) + o_\delta(1) \le \frac{2}{3}R_* + o_\gamma(1) + o_\delta(1).
\end{aligned} \tag{8.235}$$

However, this is impossible for sufficiently small γ and δ. Hence, the claim (8.217) is true. Suppose next that (8.218) failed. Then, for any $T \in (\tau_{\min}^1, T_{\max}^1)$, there exists $T' \in [T, T_{\max}^1)$ such that

$$\mathcal{K}(\widetilde{\psi}^1(T')) \le \frac{2}{3}\kappa_1(R_*). \tag{8.236}$$

Let $N_8(T')$ be the number obtained above. Then, we find from (8.37), (8.231) with $\lambda_n^1 \equiv 1$ and (8.236) that

$$\begin{aligned}\kappa_1(R_*) &\le \mathcal{K}(\psi_n(T'+t_n^1)) \\ &\le \mathcal{K}(\psi_n^1(T'+t_n^1)) + o_\gamma(1) + o_\delta(1) \le \frac{2}{3}\kappa_1(R_*) + o_\gamma(1) + o_\delta(1).\end{aligned} \tag{8.237}$$

However, for a sufficiently small γ and δ, this is a contradiction. Hence, the claim (8.218) is true.

It remains to consider the case $\tau_\infty^1 \in \mathbb{R}$. In this case, we have that $\tau_{\min}^1 < \tau_\infty^1 = -t_\infty^1 < T_{\max}^1$ (see (8.220)). Let $T' \in [-t_\infty^1, T_{\max}^1)$ be a time for which (8.234) or (8.236) holds. Then, we can derive a contradiction as well as the case $\tau_\infty^1 = -\infty$. □

LEMMA 8.0.28. *Assume* (8.127). *Then, we have* $\mathcal{M}(\widetilde{\psi}^1) = \mathcal{M}(\Phi_\omega)$.

PROOF OF LEMMA 8.0.28. We see from $\lambda_n^1 \equiv 1$ (see Lemma 8.0.27), (8.50) with $k=2$, (8.76), Lemma 8.0.26 and Lemma 8.0.7 that for any $\delta > 0$, there exists a number $N(\delta)$ such that for any $n \ge N(\delta)$,

$$\begin{aligned}\omega\delta &\ge \omega\big|\mathcal{M}(\psi_n(0)) - \mathcal{M}(e^{-it_n^1\Delta}\widetilde{u}^1)\big| - \omega\mathcal{M}(g_n^2 e^{-it_n^2\Delta}\widetilde{u}^2) - \omega\mathcal{M}(w_n^2) \\ &\ge \omega\big|\mathcal{M}(\psi_n(0)) - \mathcal{M}(e^{-it_n^1\Delta}\widetilde{u}^1)\big| - \delta \\ &\ge \omega\big|\mathcal{M}(\Phi_\omega) - \mathcal{M}(\widetilde{\psi}^1)\big| - \omega\delta - \delta.\end{aligned} \tag{8.238}$$

Since δ is an arbitrary constant, (8.238) implies the desired result. □

PROPOSITION 8.0.29. *There exists a solution* Ψ *to* (NLS) *such that*

$$\Psi \in S_{\omega,R_*,+}^{\varepsilon_*}, \tag{8.239}$$

$$\inf_{t\in[0,T_{\max})} \mathcal{K}(\Psi(t)) \ge \frac{\kappa_1(R_*)}{2}, \tag{8.240}$$

$$\mathcal{H}(\Psi) = E_*, \tag{8.241}$$

$$\|\Psi\|_{W_{p+1}([0,T_{\max}))\cap W_{2^*}([0,T_{\max}))} = \infty, \tag{8.242}$$

$$\big\{\Psi(t)\colon t\in[0,T_{\max})\big\} \text{ is precompact in } H^1(\mathbb{R}^d), \tag{8.243}$$

where $T_{\max}$ *denotes the maximal lifespan of* Ψ.

PROOF OF PROPOSITION 8.0.29. Without loss of generality, we may assume (8.127). Then, we see from Lemma 8.0.27 that $\lambda_\infty^1 = 1$ (hence $\sigma_\infty^1 = 1$ and $\widetilde{\psi}^1$ is a solution to (NLS)), and there exists $T \in I_{\max}^1$ such that

$$\inf_{t\in[T,T_{\max}^1)} \widetilde{d}_\omega(\widetilde{\psi}^1(t)) \ge \frac{R_*}{2}, \qquad \inf_{t\in[T,T_{\max}^1)} \mathcal{K}(\widetilde{\psi}^1(t)) \ge \frac{\kappa_1(R_*)}{2}. \tag{8.244}$$

We also see from (8.31), (8.33), (8.36), (8.52), (8.76) and Lemma 8.0.16 that

$$\mathcal{S}_\omega(\widetilde{\psi}^1) \le \lim_{n\to\infty} \mathcal{S}_\omega(\psi_n) \le \omega\mathcal{M}(\Phi_\omega) + E_* = m_\omega. \tag{8.245}$$

This together with Lemma 8.0.28 and (8.244) shows that $\widetilde{\psi}^1(\cdot + T) \in S_{\omega,\frac{R_*}{2}}^{\frac{\varepsilon_*}{2}}$. Furthermore, we find from Lemma 8.0.2, Lemma 8.0.4, (8.244) and Lemma 8.0.3 that

$\widetilde{\psi}^1(\cdot + T_0) \in S^{\varepsilon_*}_{\omega,R_*,+}$ for some $T_0 \ge T$. Put $\Psi(t) := \widetilde{\psi}^1(t+T_0)$. Then, the maximal lifespan of Ψ is $T_{\max}(\Psi) := T^1_{\max} - T_0$, and $\Psi \in S^{\varepsilon_*}_{\omega,R_*,+}$. We also find from Proposition 8.0.25, (1.6) and (8.244) that

$$\|\Psi\|_{W_{p+1}([0,T_{\max}(\Psi)))\cap W_{2^*}([0,T_{\max}(\Psi))} = \infty, \tag{8.246}$$

$$\mathcal{H}(\Psi) \ge \frac{1}{2} \inf_{t\in[0,T_{\max}(\Psi))} \mathcal{K}(\Psi(t)) \ge \frac{\kappa_1(R_*)}{4}. \tag{8.247}$$

Since $\mathcal{M}(\Psi) = \mathcal{M}(\widetilde{\psi}^1) = \mathcal{M}(\Phi_\omega)$, the definition of E_* (see (8.28)) together with (8.245) and (8.246) shows the property (8.241).

Now, suppose for contradiction that there existed a number $j \ge 2$ such that $\widetilde{\psi}^j$ is non-trivial. Without loss of generality, we may assume that $\widetilde{\psi}^2$ is non-trivial. Then, it follows from (8.135) and Theorem 1.0.2 that

$$\begin{cases} \mathcal{H}(\widetilde{\psi}^2) > 0 & \text{if} \quad \lambda^2_\infty = 1, \\ \mathcal{H}^\ddagger(\sigma^2_\infty \widetilde{\psi}^2) > 0 & \text{if} \quad \lambda^2_\infty = 0. \end{cases} \tag{8.248}$$

We see from (8.79), Lemma 8.0.9 and (8.248) that

$$0 < h_2 := \begin{cases} \lim_{n\to\infty} \mathcal{H}\big(g^2_n \sigma^2_n e^{-i\frac{t^2_n}{(\lambda^2_n)^2}\Delta} \widetilde{u}^2\big) = \mathcal{H}(\widetilde{\psi}^2) & \text{if} \quad \lambda^2_\infty = 1, \\ \lim_{n\to\infty} \mathcal{H}\big(g^2_n \sigma^2_n e^{-i\frac{t^2_n}{(\lambda^2_n)^2}\Delta} \widetilde{u}^2\big) = \mathcal{H}^\ddagger(\sigma^2_\infty \widetilde{\psi}^2) & \text{if} \quad \lambda^2_\infty = 0. \end{cases} \tag{8.249}$$

Furthermore, we see from (8.51), (8.36), (8.120) in Lemma 8.0.16 and (8.249) that

$$\begin{aligned} 0 &= \lim_{n\to\infty} \Big\{ \mathcal{H}(\psi_n) - \sum_{j=1}^{2} \mathcal{H}\big(g^j_n \sigma^j_n e^{-i\frac{t^j_n}{(\lambda^j_n)^2}\Delta} \widetilde{u}^j\big) - \mathcal{H}(w^2_n) \Big\} \\ &\le E_* - \lim_{n\to\infty} \mathcal{H}\big(e^{-it^1_n\Delta}\widetilde{u}^1\big) - h_2. \end{aligned} \tag{8.250}$$

This together with (8.79) shows that

$$\mathcal{H}(\Psi) = \mathcal{H}(\widetilde{\psi}^1) = \lim_{n\to\infty} \mathcal{H}\big(e^{-it^1_n\Delta}\widetilde{u}^1\big) \le E_* - h_2. \tag{8.251}$$

However, this contradicts the proved property (8.241). Thus, we have found that $\widetilde{\psi}^j \equiv 0$ for all $j \ge 2$. Then, we also see that $\widetilde{u}^j \equiv 0$ for all $j \ge 2$.

We return to the decomposition (8.43): $\psi_n(0) = e^{-it^1_n\Delta}\widetilde{u}^1 + w^1_n$. It follows from (8.50), (8.79) and $\mathcal{M}(\widetilde{\psi}^1) = \mathcal{M}(\Psi) = \mathcal{M}(\Phi_\omega) = \mathcal{M}(\psi_n)$ that

$$\lim_{n\to\infty} \|w^1_n\|_{L^2} = 0. \tag{8.252}$$

Moreover, it follows from (8.36), (8.51), (8.79) and (8.241) that

$$\lim_{n\to\infty} \mathcal{H}(w^1_n) = 0, \tag{8.253}$$

which together with (8.120) in Lemma 8.0.16 shows

(8.254)

$$\lim_{n\to\infty} \|\nabla w^1_n\|^2_{L^2} \le \frac{d}{s_p} \lim_{n\to\infty} \Big\{ \mathcal{H}(w^1_n) - \frac{2}{d(p-1)} \mathcal{K}(w^k_n) \Big\} \le \frac{d}{s_p} \lim_{n\to\infty} \mathcal{H}(w^1_n) = 0.$$

Thus, we find that

$$\lim_{n\to\infty} \|w^1_n\|_{H^1} = 0. \tag{8.255}$$

Finally, we shall prove the precompactness of $\{\Psi(t)\}$ in $H^1(\mathbb{R}^d)$. Take a sequence $\{\tau_n\}$ in $[0, T_{\max}(\Psi))$. By the continuity in time, it suffices to consider the case where $\lim_{n\to\infty} \tau_n = T_{\max}(\Psi)$. Applying the above argument to $\{\Psi(t+\tau_n)\}$, we can take a subsequence of $\{\tau_n\}$ (still denoted by the same symbol), a sequence $\{t_n\}$ in $\mathbb{R}$, a sequence $\{w_n\}$ in $H^1(\mathbb{R}^d)$ and a function $\widetilde{u} \in H^1(\mathbb{R}^d)$ such that

$$\Psi(\tau_n) = e^{-it_n\Delta}\widetilde{u} + w_n, \tag{8.256}$$

$$\lim_{n\to\infty} \|w_n\|_{H^1} = 0, \tag{8.257}$$

$$t_\infty := \lim_{n\to\infty} t_n \in \mathbb{R} \cup \{\infty\}, \tag{8.258}$$

and if $t_\infty = \infty$, then $T_{\min}(\Psi) = -\infty$. We see from (8.256) and (8.257) that

$$\lim_{n\to\infty} \|\Psi(\tau_n) - e^{-it_n\Delta}\widetilde{u}\|_{H^1} = 0. \tag{8.259}$$

Hence, in order to prove the precompactness, it suffices to show that $t_\infty \in \mathbb{R}$. Suppose for contradiction that $t_\infty = \infty$. Then, we find from Strichartz' estimate and (8.259) that

(8.260)

$$\begin{aligned}
&\lim_{n\to\infty} \|e^{it\Delta}\Psi(\tau_n)\|_{V_{2+\frac{4}{d}}((-\infty,0])\cap W_{2^*}((-\infty,0])} \\
&= \lim_{n\to\infty} \|e^{i(t+t_n)\Delta}\Psi(\tau_n)\|_{V_{2+\frac{4}{d}}((-\infty,-t_n])\cap W_{2^*}((-\infty,-t_n])} \\
&\le \lim_{n\to\infty} \|e^{i(t+t_n)\Delta}e^{-it_n\Delta}\widetilde{u}\|_{V_{2+\frac{4}{d}}((-\infty,-t_n])\cap W_{2^*}((-\infty,-t_n])} \\
&\quad + \lim_{n\to\infty} \|e^{i(t+t_n)\Delta}\{\Psi(\tau_n) - e^{-it_n\Delta}\widetilde{u}\}\|_{V_{2+\frac{4}{d}}((-\infty,-t_n])\cap W_{2^*}((-\infty,-t_n])} \\
&\lesssim \lim_{n\to\infty} \|\langle\nabla\rangle e^{it\Delta}\widetilde{u}\|_{V_{2+\frac{4}{d}}((-\infty,-t_n])\cap V_{2^*}((-\infty,-t_n])} + \lim_{n\to\infty} \|\Psi(\tau_n) - e^{-it_n\Delta}\widetilde{u}\|_{H^1} \\
&= 0.
\end{aligned}$$

Thus, the small-data theory (see Lemma D.0.1 below) shows that there exists a number $N \ge 1$ such that for any $n \ge N$,

(8.261)

$$\|\Psi\|_{W_{p+1}((-\infty,\tau_n])\cap W_{2^*}((-\infty,\tau_n])} = \|\Psi(\cdot+\tau_n)\|_{W_{p+1}((-\infty,0])\cap W_{2^*}((-\infty,0])} \lesssim 1,$$

which together with $\lim_{n\to\infty} \tau_n = T_{\max}(\Psi)$ shows

$$\|\Psi\|_{W_{p+1}((-\infty,T_{\max}(\Psi)))\cap W_{2^*}((-\infty,T_{\max}(\Psi)))} \lesssim 1. \tag{8.262}$$

However, this contradicts (8.242). Thus, we have $t_\infty \in \mathbb{R}$ and therefore the precompactness holds. □

Now, we finish the proof of Proposition 8.0.6. Under the hypothesis (8.30), we have shown the existence of solution Ψ with the properties (8.239) through (8.243) in Proposition 8.0.29. However, the same argument as the proof of Proposition 5.3 in [**18**] shows that such a solution Ψ never exists, and therefore we have arrived at a contradiction. Thus, the hypothesis (8.30) is false, and we have proved Proposition 8.0.6. □

APPENDIX A

Existence of ground state

In this section, we show the existence of ground state of (1.7) in three dimensions. We also give a proof of (1.34) at the end of this section.

PROPOSITION A.0.1. *Assume $d = 3$ and $1 < p < 5$. Then, there exists $\omega_3 > 0$ such that for any $0 < \omega < \omega_3$, a ground state of* (1.7) *exists.*

In order to prove Proposition A.0.1, we consider the following equation in $\mathbb{R}^d$:

$$v - \Delta v - \omega^{-\frac{5-p}{4}}|v|^{p-1}v - |v|^{\frac{4}{d-2}}v = 0, \qquad v \in H^1(\mathbb{R}^d) \setminus \{0\}. \tag{A.1}$$

The action associated with the equation (A.1) is

$$\widetilde{\mathcal{S}}_\omega(v) = \frac{1}{2}\|v\|_{L^2}^2 + \frac{1}{2}\|\nabla v\|_{L^2}^2 - \frac{\omega^{-\frac{5-p}{4}}}{p+1}\|v\|_{L^{p+1}}^{p+1} - \frac{1}{2^*}\|v\|_{L^{2^*}}^{2^*}. \tag{A.2}$$

It is easy to see that: if $d = 3$ and v is a solution to (A.1), then the H^1-rescaled function $\omega^{\frac{1}{4}}v(\sqrt{\omega}\cdot)$ becomes a solution to (1.7) in $\mathbb{R}^3$; and if u is a solution to (1.7) in $\mathbb{R}^3$, then $\omega^{-\frac{1}{4}}u(\cdot/\sqrt{\omega})$ becomes a solution to (A.1). Furthermore, if $d = 3$ and $\widetilde{\Phi}_\omega$ is a ground state of (A.1) (namely, $\widetilde{\Phi}_\omega$ is a solution which minimizes the action $\widetilde{\mathcal{S}}_\omega$ among the solutions to (A.1)), then for any solution u to (1.7) with $d = 3$,

$$\mathcal{S}_\omega(\omega^{\frac{1}{4}}\widetilde{\Phi}_\omega(\sqrt{\omega}\cdot)) = \widetilde{\mathcal{S}}_\omega(\widetilde{\Phi}_\omega) \le \widetilde{\mathcal{S}}_\omega(\omega^{-\frac{1}{4}}u(\cdot/\sqrt{\omega})) = \mathcal{S}_\omega(u). \tag{A.3}$$

Thus, we find that it is sufficient for Proposition A.0.1 to prove the existence of ground state of (A.1) in $\mathbb{R}^3$. An advantage to consider the equation (A.2) is that we can take into account the smallness of ω easier than the original problem (1.7) in our variational argument below.

We define

$$\widetilde{m}_\omega := \inf\left\{\widetilde{\mathcal{T}}_\omega(v) \colon v \in H^1(\mathbb{R}^3) \setminus \{0\},\ \widetilde{\mathcal{N}}_\omega(v) \le 0\right\}, \tag{A.4}$$

where

$$\widetilde{\mathcal{N}}_\omega(v) := \|v\|_{L^2}^2 + \|\nabla v\|_{L^2}^2 - \omega^{-\frac{5-p}{4}}\|v\|_{L^{p+1}}^{p+1} - \|v\|_{L^{2^*}}^{2^*}, \tag{A.5}$$

$$\widetilde{\mathcal{T}}_\omega(v) := \frac{p-1}{2(p+1)}\left\{\|v\|_{L^2}^2 + \|\nabla v\|_{L^2}^2\right\} + \frac{5-p}{6(p+1)}\|v\|_{L^{2^*}}^{2^*}. \tag{A.6}$$

It is easy to verify that

$$\widetilde{\mathcal{T}}_\omega = \widetilde{\mathcal{S}}_\omega - \frac{1}{p+1}\widetilde{\mathcal{N}}_\omega. \tag{A.7}$$

Moreover, we have the following:

LEMMA A.0.2. *Assume $d = 3$ and $1 < p < 5$. Then, we have the following:*

(i)

$$\widetilde{m}_\omega = \inf\left\{\widetilde{\mathcal{S}}_\omega(v) \colon v \in H^1(\mathbb{R}^3) \setminus \{0\},\ \widetilde{\mathcal{N}}_\omega(v) = 0\right\}. \tag{A.8}$$

(ii) *Any minimizer of the variational problem associated with* $\widetilde{m}_\omega$ *becomes a ground state of* (A.1).

Proof of Lemma A.0.2. First, we shall prove (i). Since $\widetilde{\mathcal{T}}_\omega(v) = \widetilde{\mathcal{S}}_\omega(v)$ for every $v \in H^1(\mathbb{R}^3)$ with $\widetilde{\mathcal{N}}_\omega(v) = 0$, we see that $\widetilde{m}_\omega \le \inf\{\widetilde{\mathcal{S}}_\omega(v)\colon v \in H^1(\mathbb{R}^3)\setminus\{0\},\ \widetilde{\mathcal{N}}_\omega(v) = 0\}$. In order to prove the opposite inequality, we note that for any $v_0 \in H^1(\mathbb{R}^3)\setminus\{0\}$ with $\widetilde{\mathcal{N}}_\omega(v_0) \le 0$, there exists $0 < \lambda_0 \le 1$ such that $\widetilde{\mathcal{N}}_\omega(\lambda_0 v_0) = 0$. Furthermore, this together with (A.7) shows that

(A.9)
$$\inf\{\widetilde{\mathcal{S}}_\omega(v)\colon v \in H^1(\mathbb{R}^3)\setminus\{0\},\ \widetilde{\mathcal{N}}_\omega(v) = 0\} \le \widetilde{\mathcal{S}}_\omega(\lambda_0 v_0) = \widetilde{\mathcal{T}}_\omega(\lambda_0 v_0) \le \widetilde{\mathcal{T}}_\omega(v_0).$$

Since v_0 is an arbitrary function in $H^1(\mathbb{R}^3)\setminus\{0\}$ with $\widetilde{\mathcal{N}}_\omega(v_0) \le 0$, we find that (A.8) holds.

Next, we shall prove the claim (ii). It is well known that if $\widetilde{Q}_\omega$ is a minimizer of $\inf\{\widetilde{\mathcal{S}}_\omega(v)\colon v \in H^1(\mathbb{R}^3)\setminus\{0\},\ \widetilde{\mathcal{N}}_\omega(v) = 0\}$, then $\widetilde{Q}_\omega$ is also a ground state of (A.1). Hence, we see from (A.7) and (A.8) that it suffices to prove that $\widetilde{\mathcal{N}}_\omega(\widetilde{Q}_\omega) = 0$ for all minimizer $\widetilde{Q}_\omega$ for $\widetilde{m}_\omega$. Suppose to the contrary that there exists a minimizer $\widetilde{Q}_\omega$ for $\widetilde{m}_\omega$ such that $\widetilde{\mathcal{N}}_\omega(\widetilde{Q}_\omega) < 0$. Then, we could take $0 < \lambda_0 < 1$ such that $\widetilde{\mathcal{N}}_\omega(\lambda_0\widetilde{Q}_\omega) = 0$. Furthermore, we see from the definition of $\widetilde{m}_\omega$ (see (A.4)) and $0 < \lambda_0 < 1$ that

$$\widetilde{m}_\omega \le \widetilde{\mathcal{T}}_\omega(\lambda_0\widetilde{Q}_\omega) < \widetilde{\mathcal{T}}_\omega(\widetilde{Q}_\omega) = \widetilde{m}_\omega, \tag{A.10}$$

which is a contradiction. Thus, $\widetilde{\mathcal{N}}_\omega(\widetilde{Q}_\omega) = 0$. □

A key lemma to prove the existence of the minimizer is the following:

Lemma A.0.3. *Assume $d = 3$ and $1 < p < 5$. Then, there exists $\omega_3 > 0$ such that for any $0 < \omega < \omega_3$,*

$$0 < \widetilde{m}_\omega < \frac{1}{3}\sigma^{\frac{3}{2}}. \tag{A.11}$$

Proof of Lemma A.0.3. First, we shall show that $\widetilde{m}_\omega > 0$. Let v be a non-trivial function in $H^1(\mathbb{R}^3)$ such that $\widetilde{\mathcal{N}}_\omega(v) \le 0$. Then, it follows from Sobolev's embedding that

(A.12)
$$\|v\|_{H^1}^2 = \|v\|_{L^2}^2 + \|\nabla v\|_{L^2}^2 \le \omega^{-\frac{5-p}{4}}\|v\|_{L^{p+1}}^{p+1} + \|v\|_{L^{2^*}}^{2^*} \lesssim \omega^{-\frac{5-p}{4}}\|v\|_{H^1}^{p+1} + \|v\|_{H^1}^6.$$

In particular, (A.12) implies that

$$\omega^{\frac{5-p}{4}} \lesssim \|v\|_{H^1}^{p-1} + \omega^{\frac{5-p}{4}}\|v\|_{H^1}^4. \tag{A.13}$$

Since the implicit constant of the inequality (A.13) depends only on p, we find that for any $1 < p < 5$ and any $\omega > 0$, there exists $C(\omega) > 0$ such that $C(\omega) \le \widetilde{\mathcal{T}}_\omega(v)$ for all $v \in H^1(\mathbb{R}^d)\setminus\{0\}$ with $\widetilde{\mathcal{N}}_\omega(v) \le 0$. Thus, we have proved that $\widetilde{m}_\omega > 0$.

Next, we shall show that $\widetilde{m}_\omega < \frac{1}{3}\sigma^{\frac{3}{2}}$. We define

$$W_\omega(x) := \omega^{-\frac{1}{2}}W(x/\omega) = \omega^{\frac{1}{2}}\Big(\omega^2 + \frac{|x|^2}{3}\Big)^{-\frac{1}{2}}. \tag{A.14}$$

Note that W_ω does not belong to $L^2(\mathbb{R}^3)$. Let χ be an even smooth function on $\mathbb{R}$ such that $\chi(r) = 1$ for $0 \le r \le 1$, $\chi(r) = 0$ for $r \ge 2$, and χ is non-increasing on

$[0,\infty)$. Then, we define

$$\widetilde{W}_{\omega}(x) := \chi(|x|)W_{\omega}(x). \tag{A.15}$$

We see from (1.28) that

$$\|\nabla\widetilde{W}_{\omega}\|_{L^2}^2 = \sigma^{\frac{3}{2}} + O(\omega), \tag{A.16}$$

$$\|\widetilde{W}_{\omega}\|_{L^{2^*}}^{2^*} = \sigma^{\frac{3}{2}} + O(\omega^3). \tag{A.17}$$

Moreover, we find that for any $2 \le q < 6$,

$$\|\widetilde{W}_{\omega}\|_{L^q}^q = \begin{cases} O(\omega^{\frac{6-q}{2}}) & \text{if } \ 3 < q < 6, \\ O(\omega^{\frac{3}{2}}|\log\omega|) & \text{if } \quad q = 3, \\ O(\omega^{\frac{q}{2}}) & \text{if } \ 2 \le q < 3. \end{cases} \tag{A.18}$$

Note here that there exists a unique $\tau_{\omega,0} > 0$ such that

$$\widetilde{\mathcal{N}}_{\omega}(\tau_{\omega,0}\widetilde{W}_{\omega}) = 0. \tag{A.19}$$

Furthermore, it follows from (A.16) through (A.19) that: if $2 < p < 5$, then

$$\begin{aligned} 0 &= \tau_{\omega,0}^2\|\widetilde{W}_{\omega}\|_{L^2}^2 + \tau_{\omega,0}^2\|\nabla\widetilde{W}_{\omega}\|_{L^2}^2 - \omega^{-\frac{5-p}{4}}\tau_{\omega,0}^{p+1}\|\widetilde{W}_{\omega}\|_{L^{p+1}}^{p+1} - \tau_{\omega,0}^{2^*}\|\widetilde{W}_{\omega}\|_{L^{2^*}}^{2^*} \\ &= \tau_{\omega,0}^2 O(\omega) + \tau_{\omega,0}^2\{\sigma^{\frac{3}{2}} + O(\omega)\} - \tau_{\omega,\varepsilon}^{p+1}O(\omega^{\frac{5-p}{4}}) - \tau_{\omega,0}^6\{\sigma^{\frac{3}{2}} + O(\omega^3)\}; \end{aligned} \tag{A.20}$$

if $p = 2$, then

$$0 = \tau_{\omega,0}^2 O(\omega) + \tau_{\omega,0}^2\{\sigma^{\frac{3}{2}} + O(\omega)\} - \tau_{\omega,0}^3 O(\omega^{\frac{3}{4}}|\log\omega|) - \tau_{\omega,0}^6\{\sigma^{\frac{3}{2}} + O(\omega^3)\}; \tag{A.21}$$

and if $1 < p < 2$, then

$$0 = \tau_{\omega,0}^2 O(\omega) + \tau_{\omega,0}^2\{\sigma^{\frac{3}{2}} + O(\omega)\} - \tau_{\omega,0}^{p+1}O(\omega^{\frac{p+1}{2}-\frac{5-p}{4}}) - \tau_{\omega,0}^6\{\sigma^{\frac{3}{2}} + O(\omega^3)\}. \tag{A.22}$$

Thus, we find from (A.20) through (A.22) that

$$\tau_{\omega,0}^4 = 1 + O(\omega) - \tau_{\omega,0}^{p-1}o_{\omega}(1), \tag{A.23}$$

where $o_{\omega}(1)$ means a certain function such that $\lim_{\omega\to 0} o_{\omega}(1) = 0$. Furthermore, since $0 < p-1 < 4$, this implies that

$$\lim_{\omega\to+0} \tau_{\omega,0} \ge \frac{1}{2}. \tag{A.24}$$

We introduce a function $y_{\omega}\colon (0,\infty) \to \mathbb{R}$ as

$$y_{\omega}(t) := \frac{1}{2}t^2\{\|\widetilde{W}_{\omega}\|_{L^2}^2 + \|\nabla\widetilde{W}_{\omega}\|_{L^2}^2\} - \frac{t^{2^*}}{2^*}\|\widetilde{W}_{\omega}\|_{L^{2^*}}^{2^*}. \tag{A.25}$$

It is easy to verify that the function y_{ω} attains its maximum only at the point

$$\tau_{\omega,\max} := \frac{\{\|\widetilde{W}_{\omega}\|_{L^2}^2 + \|\nabla\widetilde{W}_{\omega}\|_{L^2}^2\}^{\frac{1}{4}}}{\|\widetilde{W}_{\omega}\|_{L^{2^*}}^{\frac{3}{2}}}. \tag{A.26}$$

Note here that it follows from the definition of σ (see (1.27)), (A.16) and (A.17) that

$$\frac{\|\nabla\widetilde{W}_{\omega}\|_{L^2}^2}{\|\widetilde{W}_{\omega}\|_{L^{2^*}}^2} = \frac{\sigma^{\frac{3}{2}} + O(\omega)}{\sigma^{\frac{1}{2}} + O(\omega^3)} = \sigma + O(\omega). \tag{A.27}$$

Moreover, we see from (A.16) and (A.18) that

$$\frac{\|\widetilde{W}_\omega\|_{L^2}^2}{\|\nabla\widetilde{W}_\omega\|_{L^2}^2} = O(\omega). \tag{A.28}$$

Hence, if ω is sufficiently small, then we have

(A.29)

$$\begin{aligned} y(\tau_{\omega,\max}) &= \frac{1}{3}\Big(\frac{\|\widetilde{W}_\omega\|_{L^2}^2 + \|\nabla\widetilde{W}_\omega\|_{L^2}^2}{\|\widetilde{W}_\omega\|_{L^{2^*}}^2}\Big)^{\frac{3}{2}} = \frac{1}{3}\frac{\|\nabla\widetilde{W}_\omega\|_{L^2}^3}{\|\widetilde{W}_\omega\|_{L^{2^*}}^3}\Big(1 + \frac{\|\widetilde{W}_\omega\|_{L^2}^2}{\|\nabla\widetilde{W}_\omega\|_{L^2}^2}\Big)^{\frac{3}{2}} \\ &= \frac{1}{3}\frac{\|\nabla\widetilde{W}_\omega\|_{L^2}^3}{\|\widetilde{W}_\omega\|_{L^{2^*}}^3}\Big\{1 + \frac{3}{2}\frac{\|\widetilde{W}_\omega\|_{L^2}^2}{\|\nabla\widetilde{W}_\omega\|_{L^2}^2} + O\Big(\frac{\|\widetilde{W}_\omega\|_{L^2}^4}{\|\nabla\widetilde{W}_\omega\|_{L^2}^4}\Big)\Big\} = \frac{1}{3}\sigma^{\frac{3}{2}} + O(\omega). \end{aligned}$$

Furthermore, it follows from (A.8), (A.18), (A.19), the same computations concerning $\|\widetilde{W}_\omega\|_{L^{p+1}}^{p+1}$ as (A.20) through (A.22), and (A.24) that

$$\begin{aligned} \widetilde{m}_\omega \le \widetilde{\mathcal{S}}_\omega(\tau_{\omega,0}\widetilde{W}_\omega) &= y_\omega(\tau_{\omega,0}) - \frac{\omega^{-\frac{5-p}{4}}}{p+1}\tau_{\omega,0}^{p+1}\|\widetilde{W}_\omega\|_{L^{p+1}}^{p+1} \\ &\le y_\omega(\tau_{\omega,\max}) - c\omega^\theta = \frac{1}{3}\sigma^{\frac{3}{2}} + O(\omega) - c\omega^\theta \end{aligned} \tag{A.30}$$

for some $0 < \theta < 1$ and some $c > 0$ depending only on p. Thus, we find that if ω is sufficiently small depending only on p, then (A.11) holds. □

Now, we give a brief proof of Proposition A.0.1:

PROOF OF PROPOSITION A.0.1. As mentioned above, if $d = 3$ and $\widetilde{\Phi}_\omega$ is a ground state of (A.1), then $\omega^{\frac{1}{4}}\widetilde{\Phi}_\omega(\sqrt{\omega}\cdot)$ is one of (1.7). We also see from Lemma A.0.2 that any minimizer of the variational problem associated with $\widetilde{m}_\omega$ becomes a ground state of (A.1). Thus, it suffices to prove that for the frequency $\omega_3 > 0$ given in Lemma A.0.3 and any $0 < \omega < \omega_3$, there exists a minimizer for $\widetilde{m}_\omega$. The proof is standard: using Lemma A.0.3 and the Schwarz symmetrization, we can prove the existence of minimizer for $\widetilde{m}_\omega$ (see, e.g, Propositoin 2.1 in [**2**], and [**7**]). □

At the end of this section, we discuss the variational value m_ω defined by (1.9) in three dimensions.

PROPOSITION A.0.4. *Assume $d = 3$ and $1 + \frac{4}{3} < p < 5$. Let ω_3 be the frequency given in Proposition* A.0.1. *Then, for any $0 < \omega < \omega_3$,*

$$0 < m_\omega < \frac{1}{3}\sigma^{\frac{3}{2}}. \tag{A.31}$$

Furthermore, any ground state Q_ω of (1.7) *satisfies*

$$\mathcal{S}_\omega(Q_\omega) = m_\omega. \tag{A.32}$$

PROOF OF PROPOSITION A.0.4. First, we shall show that $m_\omega > 0$. To this end, we recall that (1.10) holds for $d = 3$ and $1 + \frac{4}{3} < p < 5$. Let u be a nontrivial function in $H^1(\mathbb{R}^3)$ such that $\mathcal{K}(u) \le 0$. Then, it follows from Sobolev's embedding that

$$\|\nabla u\|_{L^2}^2 \lesssim \|u\|_{L^2}^{\frac{5-p}{2}}\|\nabla u\|_{L^2}^{\frac{3(p-1)}{2}} + \|\nabla u\|_{L^2}^6. \tag{A.33}$$

In particular, if $\|u\|_{L^2} \ll 1$, then (A.33) implies that $1 \lesssim \|\nabla u\|_{L^2}$. Hence, if $\|u\|_{L^2} \ll 1$, then $\mathcal{I}_\omega(u) \geq \frac{p-1-\frac{4}{3}}{p-1}\|\nabla u\|_{L^2}^2 \gtrsim 1$. On the other hand, if $\|u\|_{L^2} \gtrsim 1$, then $\mathcal{I}_\omega(u) \geq \frac{\omega}{2}\|u\|_{L^2}^2 \gtrsim \omega$. Thus, we find that $\mathcal{I}_\omega(u) \gtrsim \min\{1, \omega\}$ for all $u \in H^1(\mathbb{R}^3)\backslash\{0\}$ with $\mathcal{K}(u) \leq 0$. This together with (1.10) shows $m_\omega > 0$.

Next, we shall prove (A.31). Let $\widetilde{Q}_\omega$ be a minimizer of the variational problem associated with $\widetilde{m}_\omega$ (see the proof of Proposition A.0.1 for the existence of the minimizer). Then, it follows from Lemma A.0.2 that $\widetilde{Q}_\omega$ is a ground state of (A.1). Furthermore, it is easy to verify that $\widetilde{\mathcal{N}}_\omega(\widetilde{Q}) = 0$. Put $Q_\omega := \omega^{\frac{1}{4}}\widetilde{Q}_\omega(\sqrt{\omega}\cdot)$. Then, Q_ω becomes a ground state of (1.7) and satisfies the following identity:

$$\mathcal{K}(Q_\omega) = 0. \tag{A.34}$$

Hence, we find from the definition of m_ω, (A.8) and Lemma A.0.3 that

$$m_\omega \leq \mathcal{S}_\omega(Q_\omega) = \widetilde{\mathcal{S}}_\omega(\widetilde{Q}_\omega) = \widetilde{m}_\omega < \frac{1}{3}\sigma^{\frac{3}{2}}. \tag{A.35}$$

Then, the standard argument (see, e.g., Proposition 2.1 in [**2**]) proves the existence of minimizer of the variational problem associated with m_ω. Furthermore, it is well-known that any minimizer for m_ω becomes a ground state of (1.7) (see, e.g., Proposition 1.1 in [**2**]). Thus, (A.32) holds. □

Next, we give a proof of (1.34).

PROOF OF (1.34). We find from (1.28) that the function W (see (1.26)) satisfies $\mathcal{K}^\ddagger(W) = 0$ and $\mathcal{I}^\ddagger(W) = \mathcal{H}^\ddagger(W) = \frac{1}{d}\sigma^{\frac{d}{2}}$. Hence, $m^\ddagger \leq \frac{1}{d}\sigma^{\frac{d}{2}}$ and therefore it remains to prove $\frac{1}{d}\sigma^{\frac{d}{2}} \leq m^\ddagger$. Let $u \in H^1(\mathbb{R}^d) \setminus \{0\}$ satisfy $\mathcal{K}^\ddagger(u) \leq 0$. Then, we can take $0 < \lambda_0 \leq 1$ such that $\mathcal{K}^\ddagger(\lambda_0 u) = 0$. Furthermore, it follows from the definition of σ (see (1.27)) and $\mathcal{K}^\ddagger(\lambda_0 u) = \lambda_0^2\|\nabla u\|_{L^2}^2 - \lambda_0^{2^*}\|u\|_{L^{2^*}}^{2^*} = 0$ that

$$\sigma \leq \frac{\|\nabla u\|_{L^2}^2}{\|u\|_{L^{2^*}}^2} \leq \lambda_0^{\frac{4}{d-2}}\|u\|_{L^{2^*}}^{\frac{4}{d-2}}. \tag{A.36}$$

Raising the both sides of (A.36) to the power of $\frac{d}{2}$, and using $\mathcal{K}^\ddagger(\lambda_0 u) = 0$ and $\lambda_0 \leq 1$, we see that

$$\begin{aligned} \frac{1}{d}\sigma^{\frac{d}{2}} &\leq \frac{1}{d}\|\lambda_0 u\|_{L^{2^*}}^{2^*} = \mathcal{H}^\ddagger(\lambda_0 u) \\ &= \mathcal{H}^\ddagger(\lambda_0 u) - \frac{2}{d(p-1)}\mathcal{K}^\ddagger(\lambda_0 u) = \mathcal{I}^\ddagger(\lambda_0 u) \leq \mathcal{I}^\ddagger(u). \end{aligned} \tag{A.37}$$

Since u is arbitrary, we find that $\frac{1}{d}\sigma^{\frac{d}{2}} \leq m^\ddagger$. Thus, we have completed the proof. □

APPENDIX B

Fundamental properties of the linearized operators

In this section, we discuss fundamental properties of the operators $L_{\omega,+}$ and $L_{\omega,-}$ defined by (4.4) and (4.5), respectively. Throughout this section, we assume that $d \geq 3$, $1+\frac{4}{d} < p < 2^*-1$ and $0 < \omega < \omega_1$, where ω_1 is the frequency given by Proposition 2.0.4.

First, we can easily verify that for any $f \in H^1(\mathbb{R}^d)$,

(B.1)
$$\left|\langle L_{\omega,+}f, f\rangle_{H^{-1},H^1}\right| + \left|\langle L_{\omega,-}f, f\rangle_{H^{-1},H^1}\right| \lesssim \left(\omega + 1 + \|\Phi_\omega\|_{L^{p+1}}^{p-1} + \|\Phi_\omega\|_{L^{2^*}}^{\frac{4}{d-2}}\right)\|f\|_{H^1}^2,$$

where the implicit constant is independent of ω. Moreover, we see from (4.3) and (4.6) that for any $u, v \in H^1_{real}(\mathbb{R}^d)$,

$$\left[\mathcal{S}_\omega''(\Phi_\omega)u\right]v = \langle L_{\omega,+}\Re[u] + iL_{\omega,-}\Im[u], v\rangle_{H^{-1},H^1}. \tag{B.2}$$

LEMMA B.0.1. *The operator $L_{\omega,-}$ is non-negative and* $\operatorname{Ker} L_{\omega,-} = \operatorname{span}\{\Phi_\omega\}$*; in particular, there exists $\delta_\omega > 0$ depending on ω such that for any $u \in H^1(\mathbb{R}^d)$ with $(u, \Phi_\omega)_{L^2_{real}} = 0$,*

$$\langle L_{\omega,-}u, u\rangle_{H^{-1},H^1} \geq \delta_\omega \|u\|_{H^1}^2. \tag{B.3}$$

PROOF OF LEMMA B.0.1. We can prove the claim in a way similar to [**30**]. □

Differentiating the equation (1.7) for Φ_ω with respect to x_j, $1 \leq j \leq d$, we have

$$0 = \partial_j(\omega\Phi_\omega - \Delta\Phi_\omega - \Phi_\omega^p - \Phi_\omega^{2^*-1}) = L_{\omega,+}\partial_j\Phi_\omega. \tag{B.4}$$

LEMMA B.0.2. *There exists $\omega_3 > 0$ such that for any $\omega \in (0, \omega_3)$, we have*

$$\operatorname{Ker} L_{\omega,+} = \operatorname{span}\{\partial_1\Phi_\omega, \dots, \partial_d\Phi_\omega\}. \tag{B.5}$$

PROOF OF LEMMA B.0.2. We can prove this lemma by regarding the operator $L_{\omega,+}$ as a perturbation of $L_\omega^\dagger$ (see [**13**] and Lemma 2.0.5). □

Now, we introduce the functional $\mathcal{N}_\omega$ as

$$\mathcal{N}_\omega(u) := \omega\|u\|_{L^2}^2 + \|\nabla u\|_{L^2}^2 - \|u\|_{L^{p+1}}^{p+1} - \|u\|_{L^{2^*}}^{2^*}. \tag{B.6}$$

Then, we can verify that

$$m_\omega = \inf\left\{\mathcal{S}_\omega(u) \colon u \in H^1(\mathbb{R}^d) \setminus \{0\},\ \mathcal{N}_\omega(u) = 0\right\}. \tag{B.7}$$

LEMMA B.0.3. *For any $u \in H^1(\mathbb{R}^d)$ satisfying*

$$\left(\, (p-1)\Phi_\omega^p + (2^*-2)\Phi_\omega^{2^*-1},\ u\,\right)_{L^2_{real}} = 0, \tag{B.8}$$

we have

$$\left\langle L_{\omega,+}u, u\right\rangle_{H^{-1},H^1} \geq 0. \tag{B.9}$$

PROOF OF LEMMA B.0.3. The proof is similar to Lemma 2.3 in [**24**]. □

LEMMA B.0.4. *As an operator in* $L^2_{rad}(\mathbb{R}^d)$, $L_{\omega,+}$ *has only one negative eigenvalue which is non-degenerate and* 0 *is not an eigenvalue.*

PROOF OF LEMMA B.0.4. We can prove this lemma in a way similar to [**30**] and [**13**]. □

LEMMA B.0.5. *Let* ω_2 *be the frequency given by Proposition* 4.0.1. *Then, there exists* $\omega_0 \in (0, \omega_2)$ *such that the following holds for all* $0 < \omega < \omega_0$: *Let* $\mu > 0$ *be a positive eigenvalue of* $i\mathcal{L}_\omega$ *as an operator in* $L^2_{real}(\mathbb{R}^d)$, *and let* $\mathfrak{U}_+$ *be a corresponding eigenfunction. Then, we have:*
(i) *For any non-trivial, real-valued radial function* $g \in H^1(\mathbb{R}^d)$ *with* $(g, f_2)_{L^2_{real}} = 0$,

$$\langle L_{\omega,+}g, g\rangle_{H^{-1},H^1} \sim \|g\|^2_{H^1}, \tag{B.10}$$

where the implicit constant may depend on ω.

(ii) *For any non-trivial, real-valued radial function* $g \in H^1(\mathbb{R}^d)$ *with* $(g, \partial_\omega \Phi_\omega)_{L^2_{real}} = 0$,

$$\langle L_{\omega,-}g, g\rangle_{H^{-1},H^1} \sim \|g\|^2_{H^1}, \tag{B.11}$$

where the implicit constant may depend on ω.

PROOF OF LEMMA B.0.5. We prove the claim (i).

First, we shall show that for any $0 < \omega < \omega_2$ and any nontrivial, radial real-valued function $g \in H^1(\mathbb{R}^d)$ satisfying $(g, f_2)_{L^2_{real}} = 0$,

$$\langle L_{\omega,+}g, g\rangle_{H^{-1},H^1} > 0. \tag{B.12}$$

Suppose for contradiction that there exists $\omega > 0$ and a nontrivial radial real-valued function $g_- \in H^1(\mathbb{R}^d)$ such that

$$(g_-, f_2)_{L^2_{real}} = 0, \tag{B.13}$$

$$\langle L_{\omega,+}g_-, g_-\rangle_{H^{-1},H^1} \le 0. \tag{B.14}$$

Since $L_{\omega,+}$ is self-adjoint in $L^2_{real}(\mathbb{R}^d)$, we see from (4.109) and (B.13) that

$$\langle L_{\omega,+}g_-, f_1\rangle_{H^{-1},H^1} = \big(g_-, L_{\omega,+}f_1\big)_{L^2_{real}} = -\mu\big(g_-, f_2\big)_{L^2_{real}} = 0. \tag{B.15}$$

Moreover, it follows from (4.109) and (4.114) that

$$\langle L_{\omega,+}f_1, f_1\rangle_{H^{-1},H^1} = -\big(\mu f_2, f_1\big)_{L^2_{real}} = -\mu\big(f_1, f_2\big)_{L^2_{real}} < 0. \tag{B.16}$$

Note here that (B.15) and (B.16) show that g_- and f_1 are linearly independent in $L^2_{real}(\mathbb{R}^d)$. We see from the hypothesis (B.14) and (B.15) that for any $a, b \in \mathbb{R}$,

(B.17)

$$\begin{aligned}
&\langle L_{\omega,+}\left(af_1 + b\frac{g_-}{\|g_-\|_{L^2}}\right), \left(af_1 + b\frac{g_-}{\|g_-\|_{L^2}}\right)\rangle_{H^{-1},H^1} \\
&= a^2\big(L_{\omega,+}f_1, f_1\big)_{L^2_{real}} + \frac{b^2}{\|g_-\|^2_{L^2}}\langle L_{\omega,+}g_-, g_-\rangle_{H^{-1},H^1} + \frac{2ab}{\|g_-\|_{L^2}}\langle L_{\omega,+}g_-, f_1\rangle_{H^{-1},H^1} \\
&\le a^2\big(L_{\omega,+}f_1, f_1\big)_{L^2_{real}} \le 0.
\end{aligned}$$

Put

$$e_1 := \frac{g_-}{\|g_-\|_{L^2}}, \quad e_2 := \frac{f_1 - (f_1, e_1)_{L^2_{real}} e_1}{\big\|f_1 - (f_1, e_1)_{L^2_{real}} e_1\big\|_{L^2}}. \tag{B.18}$$

Then, it is clear that $\|e_1\|_{L^2} = \|e_2\|_{L^2} = 1$ and $(e_1, e_2)_{L^2_{real}} = 0$. Moreover, we find from (B.17) that for any $\alpha, \beta \in \mathbb{R}$,

$$\langle L_{\omega,+}(\alpha e_1 + \beta e_2),\, \alpha e_1 + \beta e_2 \rangle_{H^{-1},H^1} \le 0. \tag{B.19}$$

Since Weyl's essential spectrum theorem shows $\sigma_{ess}(L_{\omega,+}) = \sigma_{ess}(-\Delta + \omega^2) = [\omega^2, \infty)$, the min-max theorem (see Theorem 12.1 in [**23**]) together with (B.19) shows that the second eigenvalue of $L_{\omega,+}$ is non-positive. However, this contradicts the fact that $L_{\omega,+}$ has only one non-positive eigenvalue as an operator in $L^2_{rad}(\mathbb{R}^d)$ (see Lemma B.0.4). Thus, we have proved the claim (B.12).

Now, we are in a position to prove the claim (i). It follows from (B.1) and (2.9) that there exists $\omega_0 > 0$ such that for any $\omega \in (0, \omega_0)$ and any $g \in H^1(\mathbb{R}^d)$,

$$\langle L_{\omega,+} g, g \rangle_{H^{-1},H^1} \lesssim \left(\omega + 1 + \omega^{\frac{p-1}{p+1}(1-s_p)} \|U\|_{L_{p+1}}^{p-1} + \omega^{\frac{4}{(d-2)(p-1)}-1} \|U\|_{L_{2^*}}^{2^*-2}\right) \|g\|_{H^1}^2. \tag{B.20}$$

Thus, it suffices to show that

$$\inf_{\substack{g \in H^1_{rad}(\mathbb{R}^d,\mathbb{R}) \\ \|g\|_{H^1}=1 \\ (g,f_2)_{L^2_{real}}=0}} \langle L_{\omega,+} g,\, g \rangle_{H^{-1},H^1} > 0, \tag{B.21}$$

where $H^1_{rad}(\mathbb{R}^d, \mathbb{R})$ denotes the space of radial real-valued functions in $H^1(\mathbb{R}^d)$. Suppose for a contradiction that (B.21) failed. Then, we could take a sequence $\{g_n\}$ of radial real-valued functions in $H^1(\mathbb{R}^d)$ such that

$$\|g_n\|_{H^1} \equiv 1, \tag{B.22}$$

$$(g_n, f_2)_{L^2_{real}} \equiv 0, \tag{B.23}$$

$$\lim_{n\to\infty} \langle L_{\omega,+} g_n, g_n \rangle_{H^{-1},H^1} = 0. \tag{B.24}$$

Here, passing to some subsequence, we may assume that there exists a radial real-valued function $g_\infty \in H^1(\mathbb{R}^d)$ such that

$$\lim_{n\to\infty} g_n = g_\infty \qquad \text{weakly in } H^1(\mathbb{R}^d). \tag{B.25}$$

If $g_\infty = 0$, then we see from (B.24) and (B.25) that for any $\omega \in (0, 1)$,

$$\begin{aligned} 0 &= \lim_{n\to\infty} \langle L_{\omega,+} g_n, g_n \rangle_{H^{-1},H^1} = \lim_{n\to\infty} \left(L_{\omega,+} g_n, g_n\right)_{L^2_{real}} \\ &= \lim_{n\to\infty} \left\{ \omega \int_{\mathbb{R}^d} |g_n|^2 + \int_{\mathbb{R}^d} |\nabla g_n|^2 - p \int_{\mathbb{R}^d} \Phi_\omega^{p-1} |g_n|^2 - (2^*-1) \int_{\mathbb{R}^d} \Phi_\omega^{\frac{4}{d-2}} |g_n|^2 \right\} \\ &\ge \omega - \lim_{n\to\infty} \left\{ p \int_{\mathbb{R}^d} \Phi_\omega^{p-1} |g_n|^2 + (2^*-1) \int_{\mathbb{R}^d} \Phi_\omega^{\frac{4}{d-2}} |g_n|^2 \right\} \\ &= \omega, \end{aligned} \tag{B.26}$$

which is a contradiction. Thus, we have found that g_∞ is nontrivial. Then, it follows from (B.24), the lower semicontinuity and (B.12) that

$$0 = \lim_{n\to\infty} \langle L_{\omega,+} g_n, g_n \rangle_{H^{-1},H^1} \ge \langle L_{\omega,+} g_\infty, g_\infty \rangle_{H^{-1},H^1} > 0. \tag{B.27}$$

This absurd conclusion comes from the hypothesis that (B.21) failed. Thus, we have prove the claim (i).

Next, we prove the claim (ii). Let $g \in H^1(\mathbb{R}^d)$ be a non-trivial, real-valued, radial function with $(g, \partial_\omega \Phi_\omega)_{L^2_{real}} = 0$. We write g in the form

$$g = a\Phi_\omega + h, \tag{B.28}$$

where h satisfies $(h, \Phi_\omega)_{L^2_{real}} = 0$. Note that the condition $(g, \partial_\omega \Phi_\omega)_{L^2_{real}} = 0$ implies that

$$a = -\frac{\big(h, \partial_\omega \Phi_\omega\big)_{L^2_{real}}}{\big(\Phi_\omega, \partial_\omega \Phi_\omega\big)_{L^2_{real}}}. \tag{B.29}$$

Hence, we find from $(h, \Phi_\omega)_{L^2_{real}} = 0$ that

(B.30)

$$\begin{aligned}
&\|g\|_{H^1}^2 = \big(g, g\big)_{L^2_{real}} + \big(\nabla g, \nabla g\big)_{L^2_{real}} \\
&= a^2 \|\Phi_\omega\|_{H^1}^2 + \|h\|_{H^1}^2 - 2a\big(\nabla \Phi_\omega, \nabla h\big)_{L^2_{real}} \\
&\le \frac{\|\partial_\omega \Phi_\omega\|_{L^2}^2 \|h\|_{L^2}^2}{\big|(\Phi_\omega, \partial_\omega \Phi_\omega)_{L^2_{real}}\big|^2} \|\Phi_\omega\|_{H^1}^2 + \|h\|_{H^1}^2 + \frac{2\|\partial_\omega \Phi_\omega\|_{L^2} \|h\|_{L^2}}{\big|(\Phi_\omega, \partial_\omega \Phi_\omega)_{L^2_{real}}\big|} \|\nabla \Phi_\omega\|_{L^2} \|\nabla h\|_{L^2} \\
&\lesssim \|h\|_{H^1}^2.
\end{aligned}$$

Moreover, we see from $L_{\omega,-}\Phi_\omega = 0$ and Lemma B.0.1, that

$$\begin{aligned}
\langle L_{\omega,-} g, g\rangle_{H^{-1},H^1} &= \langle L_{\omega,-}(a\Phi_\omega + h),\, a\Phi_\omega + h\rangle_{H^{-1},H^1} \\
&= \langle L_{\omega,-} h, h\rangle_{H^{-1},H^1} \gtrsim \|h\|_{H^1}^2.
\end{aligned} \tag{B.31}$$

Putting (B.30) and (B.31) together, we obtain

$$\|g\|_{H^1}^2 \lesssim \langle L_{\omega,-} g, g\rangle_{H^{-1},H^1}. \tag{B.32}$$

The opposite relation follows from (B.1). Hence, we have completed the proof. □

APPENDIX C

Inequalities for the radial functions

We see from the fundamental theorem of calculus and Hardy's inequality that for any $d \geq 3$ and any radial function $g \in H^1(\mathbb{R}^d)$,

$$\sup_{x\in\mathbb{R}^d} \{|x|^{d-2}|g(x)|^2\} = \sup_{x\in\mathbb{R}^d} \int_{|x|}^{\infty} \frac{d}{dr}\Big\{ -r^{d-2}|g(r)|^2\Big\}\, dr$$

$$\lesssim \int_0^\infty r^{\frac{d-3}{2}}|g(r)|r^{\frac{d-1}{2}}|g'(r)|\, dr \tag{C.1}$$

$$\leq \Big(\int_0^\infty |g(r)|^2 r^{d-3}\, dr\Big)^{\frac{1}{2}} \Big(\int_0^\infty |g'(r)|^2 r^{d-1}\, dr\Big)^{\frac{1}{2}} \lesssim \|\nabla g\|_{L^2(\mathbb{R}^d)}^2.$$

Similarly, we can verify that for any $d \geq 3$ and any radial function $g \in H^1(\mathbb{R}^d)$,

$$\sup_{x\in\mathbb{R}^d} \{|x|^{d-1}|g(x)|^2\} \lesssim \int_0^\infty r^{\frac{d-1}{2}}|g(r)|r^{\frac{d-1}{2}}|g'(r)|\, dr \lesssim \|g\|_{L^2(\mathbb{R}^d)}\|\nabla g\|_{L^2(\mathbb{R}^d)}. \tag{C.2}$$

APPENDIX D

Small-data theory

We record standard small-data theories for (NLS) and (1.19) (see Lemma E.0.1 below and Corollary 3.9 in [**22**]):

LEMMA D.0.1. (i) *Assume $d \geq 3$ and $1 + \frac{4}{d} < p < 2^* - 1$. Then, for any $A > 0$, there exists $\delta(A) > 0$ with the following property: for any $t_0 \in \mathbb{R}$ and any $\psi_0 \in H^1(\mathbb{R}^d)$ satisfying*

$$\|\langle\nabla\rangle\psi_0\|_{L^2} \leq A, \tag{D.1}$$

$$\|e^{i(t-t_0)\Delta}\psi_0\|_{W_{2+\frac{4}{d}}(\mathbb{R})\cap W_{2^*}(\mathbb{R})} \leq \delta(A), \tag{D.2}$$

there exists a unique strong H^1-solution ψ to (NLS) *with $\psi(t_0) = \psi_0$. Furthermore, the maximal existence-interval of ψ is the whole of $\mathbb{R}$, and ψ satisfies*

$$\|\langle\nabla\rangle\psi\|_{St(\mathbb{R})} \lesssim \|\langle\nabla\rangle\psi_0\|_{L^2}. \tag{D.3}$$

(ii) *Assume $d \geq 3$. Then, there exists $\delta_0 > 0$ with the following property: for any $t_0 \in \mathbb{R}$ and any $\psi_0 \in \dot{H}^1(\mathbb{R}^d)$ satisfying*

$$\|\nabla\psi_0\|_{L^2} \leq \delta_0, \tag{D.4}$$

there exists a unique strong $\dot{H}^1$-solution ψ to (1.19) *with $\psi(t_0) = \psi_0$. Furthermore, the maximal existence-interval of ψ is the whole of $\mathbb{R}$ and*

$$\|\nabla\psi\|_{St(\mathbb{R})} \lesssim \|\nabla\psi_0\|_{L^2}. \tag{D.5}$$

We can control the norm of the full Strichartz space by a few particular norms:

LEMMA D.0.2. *Assume $d \geq 3$, and let $A_1, A_2, A_3 > 0$. Then, there exists a constant $C(A_1, A_2, A_3) > 0$ with the following property:*
(i) *for any interval I and any space-time function u satisfying*

$$\|u\|_{L^\infty(I,H^1)} \leq A_1, \tag{D.6}$$

$$\|u\|_{W_{p+1}(I)\cap W_{2^*}(I)} \leq A_2, \tag{D.7}$$

$$\big\|\langle\nabla\rangle e[u]\big\|_{L_{t,x}^{\frac{2(d+2)}{d+4}}(I)} \leq A_3, \tag{D.8}$$

we have

$$\|\langle\nabla\rangle u\|_{St(I)} \leq C(A_1, A_2, A_3). \tag{D.9}$$

(ii) *for any interval I and any space-time function u satisfying*

$$\|\nabla u\|_{L^\infty(I,L^2)} \leq A_1, \tag{D.10}$$

$$\|u\|_{W_{2^*}(I)} \leq A_2, \tag{D.11}$$

$$\left\|\nabla e^{\ddagger}[u]\right\|_{L_{t,x}^{\frac{2(d+2)}{d+4}}(I)} \leq A_3, \tag{D.12}$$

we have

$$\|\nabla u\|_{St(I)} \leq C(A_1, A_2, A_3). \tag{D.13}$$

PROOF OF LEMMA D.0.2. See the proof of Lemma 4.1 in [**3**]. □

APPENDIX E

Long-time perturbation theory

In this section, we give the long-time perturbation theory for a general nonlinear Schrödinger equation including (NLS) with $1 + \frac{4}{d} \le p < 2^* - 1$:

$$i\frac{\partial \psi}{\partial t} + \Delta\psi + f(\psi) = 0, \tag{E.1}$$

where $f\colon \mathbb{C} \to \mathbb{C}$ is a continuously differentiable function in the real-sense.

Fix an even, smooth cut-off function χ defined on $\mathbb{R}$ such that $\chi(r) = 1$ if $0 \le r \le 1$ and $\chi(r) = 0$ if $r \ge 2$, and define

$$f_{\le 1}(z) := \chi(|z|)f(z), \tag{E.2}$$

$$f_{>1}(z) := f(z) - f_{\le 1}(z). \tag{E.3}$$

Then, our assumptions about the nonlinearity f are the followings:

(1)

$$f(0) = \frac{\partial f}{\partial z}(0) = \frac{\partial f}{\partial \bar{z}}(0) = 0. \tag{E.4}$$

(2) There exists a constant $C_{\le 1} > 0$ such that for any $z_1, z_2 \in \mathbb{C}$,

$$\begin{aligned} &\Big|\frac{\partial f_{\le 1}}{\partial z}(z_1) - \frac{\partial f_{\le 1}}{\partial z}(z_2)\Big| + \Big|\frac{\partial f_{\le 1}}{\partial \bar{z}}(z_1) - \frac{\partial f_{\le 1}}{\partial \bar{z}}(z_2)\Big| \\ &\le C_{\le 1}\begin{cases} |z_1 - z_2|\min\left\{10, (|z_1| + |z_2|)^{\frac{4}{d}-1}\right\} & \text{if} \quad d = 3, \\ |z_1 - z_2|^{\frac{4}{d}} & \text{if} \quad d \ge 4. \end{cases} \end{aligned} \tag{E.5}$$

(3) There exists a constant $C_{>1} > 0$ such that for any $z_1, z_2 \in \mathbb{C}$,

$$\begin{aligned} &\Big|\frac{\partial f_{>1}}{\partial z}(z_1) - \frac{\partial f_{>1}}{\partial z}(z_2)\Big| + \Big|\frac{\partial f_{>1}}{\partial \bar{z}}(z_1) - \frac{\partial f_{>1}}{\partial \bar{z}}(z_2)\Big| \\ &\le C_{>1}\begin{cases} |z_1 - z_2|(|z_1| + |z_2|)^{\frac{4}{d-2}-1} & \text{if} \quad 3 \le d \le 5, \\ |z_1 - z_2|^{\frac{4}{d-2}} & \text{if} \quad d \ge 6. \end{cases} \end{aligned} \tag{E.6}$$

Throughout this section, we allow the implicit constants to depend on $C_{\le 1}$ and $C_{>1}$, so that the assumptions (E.4) through (E.6) imply that

$$|f_{\le 1}(z)| \lesssim |z|^{1+\frac{4}{d}}, \tag{E.7}$$

$$|f_{>1}(z)| \lesssim |z|^{1+\frac{4}{d-2}}, \tag{E.8}$$

$$\left|\frac{\partial f_{\le 1}}{\partial z}(z)\right| + \left|\frac{\partial f_{\le 1}}{\partial \bar{z}}(z)\right| \lesssim |z|^{\frac{4}{d}}, \tag{E.9}$$

$$\left|\frac{\partial f_{>1}}{\partial z}(z)\right| + \left|\frac{\partial f_{>1}}{\partial \bar{z}}(z)\right| \lesssim |z|^{\frac{4}{d-2}} \tag{E.10}$$

for any $z \in \mathbb{C}$.

Now, we state the long-time perturbation theory for the equation (E.1):

LEMMA E.0.1. *Let $d \ge 3$, I an interval, $\psi \in C(I, H^1(\mathbb{R}^d))$ a solution to* (E.1), *and let u be a function in $C(I, H^1(\mathbb{R}^d))$. Assume that*

$$\|u\|_{L_t^\infty H_x^1(I)} \le A, \tag{E.11}$$

$$\|u\|_{V_{2+\frac{4}{d}}(I)\cap W_{2^*}(I)} \le B, \tag{E.12}$$

$$\|\psi(t_0) - u(t_0)\|_{H^1} \le \gamma \tag{E.13}$$

for some $t_0 \in I$ and some constants $A > 0$, $B > 0$ and $\gamma > 0$. Then, there exist constants $\delta_0(A, B, \gamma) > 0$ and $C(A, B, \gamma) > 0$ such that if

$$\|e^{i(t-t_0)\Delta}\{\psi(t_0) - u(t_0)\}\|_{V_{2+\frac{4}{d}}(I)\cap W_{2^*}(I)} \le \delta, \tag{E.14}$$

$$\|\langle\nabla\rangle e(u)\|_{L_{t,x}^{\frac{2(d+2)}{d+4}}(I)} \le \delta \tag{E.15}$$

for some $0 < \delta < \delta_0(A, B, \gamma)$, then we have

$$\|\psi - u\|_{V_{2+\frac{4}{d}}(I)\cap W_{2^*}(I)} \le C(A, B, \gamma)\delta^{\frac{1}{d(d+2)}}, \tag{E.16}$$

$$\|\langle\nabla\rangle\{\psi - u\}\|_{St(I)} \le C(A, B, \gamma)\gamma, \tag{E.17}$$

$$\|\langle\nabla\rangle\psi\|_{St(I)} \le C(A, B, \gamma). \tag{E.18}$$

Here, the constant $C(A, B, \gamma)$ is non-decreasing with respect to each of the parameters A, B and γ.

We can prove Lemma E.0.1 in a way similar to the proof of Theorem 3.8 in [**22**]. The main difference of the proof is that we need another exotic Strichartz estimate. For example, the following exotic Strichartz norms work well:

$$\|u\|_{ES(I)} := \|\langle\nabla\rangle^{\frac{4}{d+2}} u\|_{L_t^{\frac{d(d+2)}{2(d-2)}} L_x^{\frac{2d^2(d+2)}{d^3-4d+16}}(I)} + \|u\|_{L_t^{\frac{d+2}{2}} L_x^{\frac{2d(d+2)}{(d-2)(d+4)}}(I)}, \tag{E.19}$$

$$\|u\|_{ES^*(I)} := \|\langle\nabla\rangle^{\frac{4}{d+2}} u\|_{L_t^{\frac{d}{2}} L_x^{\frac{2d^2(d+2)}{d^3+4d^2+4d-16}}(I)} + \|u\|_{L_t^{\frac{d(d+2)}{2(d+4)}} L_x^{\frac{2d^2(d+2)}{(d+4)^2(d-2)}}(I)}, \tag{E.20}$$

where I is an interval. In particular, we have the following exotic Strichartz estimate.

LEMMA E.0.2 (Exotic Strichartz estimate). *Let $d \geq 3$. Then, we have*

$$\|u - e^{i(t-t_0)\Delta}u(t_0)\|_{ES(I)} \lesssim \|i\frac{\partial u}{\partial t} + \Delta u - g\|_{ES^*(I)} + \|\langle\nabla\rangle g\|_{L_{t,x}^{\frac{2(d+2)}{d+4}}(I)} \tag{E.21}$$

for any interval I, any $t_0 \in I$, and any functions u and g defined on $\mathbb{R}^d \times I$.

Moreover, we can derive an analogue of Lemma 3.11 in [**22**]:

LEMMA E.0.3. *Let $d \geq 3$. Then, we have*

$$\|u\|_{ES(I)} \lesssim \|u\|_{L_{t,x}^{\frac{2(d+2)}{d-2}}(I)}^{\frac{1}{d+2}} \|\nabla u\|_{St(I)}^{\frac{d+1}{d+2}} + \|u\|_{L_{t,x}^{\frac{2(d+2)}{d}}(I)}^{\frac{16(d-2)}{d^2(d+2)}} \|\langle\nabla\rangle u\|_{St(I)}^{\frac{d^3+2d^2-16d+32}{d^2(d+2)}} \tag{E.22}$$

for any interval I and any function u on $\mathbb{R}^d \times I$. In particular, we have

$$\|u\|_{ES(I)} \lesssim \|\langle\nabla\rangle u\|_{St(I)}. \tag{E.23}$$

Moreover, there exist constants $\theta_0, \theta_1 \in (0, 1]$ depending only on d such that

$$\|u\|_{L_{t,x}^{\frac{2(d+2)}{d}}(I)} \leq \|u\|_{St(I)}^{1-\theta_0} \|u\|_{ES(I)}^{\theta_0}, \tag{E.24}$$

$$\|u\|_{L_{t,x}^{\frac{2(d+2)}{d-2}}(I)} \leq \|\nabla u\|_{St(I)}^{1-\theta_1} \|u\|_{ES(I)}^{\theta_1} \tag{E.25}$$

for any interval I and any function u on $\mathbb{R}^d \times I$.

APPENDIX F

Table of notation

Symbols	Descriptions or equation numbers
2^*	(1.1)
$\mathcal{M}$	(1.2)
$\mathcal{H}, \mathcal{H}^\dagger, \mathcal{H}^\ddagger,$	(1.3), (1.41), (1.31)
$\mathcal{S}_\omega, \mathcal{S}_\omega^\dagger, \mathcal{S}_\omega^j$	(1.8), (1.46), (8.211)
$\mathcal{K}, \mathcal{K}^\dagger, \mathcal{K}^\ddagger$	(1.5), (1.43), (1.31)
$\mathcal{I}_\omega, \mathcal{I}^\ddagger, \mathcal{I}_\omega^j$	(1.11), (1.30), (8.211)
$\mathcal{J}_\omega$	(1.12)
Φ_ω, U	positive radial ground states of (1.7) and (1.22)
$\mathcal{O}(\Phi_\omega)$	(1.15)
W	(1.26)
$m_\omega, \sigma, m^\ddagger, \nu_\omega$	(1.9), (1.27), (1.29), (4.17)
$\widetilde{PW}_\omega, PW_\omega, PW_{\omega,+}, PW_{\omega,-}$	(1.16), (1.48), (1.50), (1.51)
$PW^\dagger, PW_\omega^\dagger, PW^{\dagger,\varepsilon}$	(1.40), (1.44), (1.47)
$PW_+^\ddagger$	(1.37)
T_ω	(1.24)
s_p	(1.42)
$\mathcal{L}_\omega, L_{\omega,+}, L_{\omega,-}$	(4.2), (4.4), (4.5)
μ	positive eigen-value of $-i\mathcal{L}_\omega$
$\mathcal{U}_\pm$	eigen-function of $-i\mathcal{L}_\omega$ corresponding to $\pm\mu$, (4.43)
$\lambda_\pm(t)$	(4.50)
$\Gamma(t)$	(4.65)
$\\|\eta(t)\\|_E$	(4.70)
$\lambda_1(t)$, $\lambda_2(t)$	(4.88)
f_1, f_2	(4.107)
$d_\omega(\psi(t)), \widetilde{d}_\omega(\psi(t))$	(4.85), (6.15)
$\delta_E(\omega), \delta_0(\omega), \delta_X, \delta_S, \delta_*$	Proposition 4.0.4, Lemma 5.0.1,
$\delta_E(\omega), \delta_0(\omega), \delta_X, \delta_S, \delta_*$	Lemma 5.0.2, (7.7), Theorem 7.0.1
T_X	(5.12)
$\kappa_1(\delta)$	Lemma 6.0.3
$A_\omega^\varepsilon, S_{\omega,R}^\varepsilon, S_{\omega,R,\pm}^\varepsilon$	(3.1), (8.1), (8.19)
$W^j(I)$	(8.73)
G_n^j, g_n^j	(8.44), (8.45)
$\sigma_n^j, \sigma_\infty^j$	(8.47), (8.56)
$x_n^j, t_n^j, \lambda_n^j$	(8.39), (8.40), (8.41)
w_n^k, $\widetilde{u}^j$	(8.43), (8.42)
$\widetilde{\psi}^j$, ψ_n^j	(8.72), (8.80)
ψ_n^{k-app}	(8.168)

Bibliography

[1] Takafumi Akahori and Hayato Nawa, *Blowup and scattering problems for the nonlinear Schrödinger equations*, Kyoto J. Math. **53** (2013), no. 3, 629–672, DOI 10.1215/21562261-2265914. MR3102564

[2] Takafumi Akahori, Slim Ibrahim, Hiroaki Kikuchi, and Hayato Nawa, *Existence of a ground state and blow-up problem for a nonlinear Schrödinger equation with critical growth*, Differential Integral Equations **25** (2012), no. 3-4, 383–402. MR2917888

[3] Takafumi Akahori, Slim Ibrahim, Hiroaki Kikuchi, and Hayato Nawa, *Existence of a ground state and scattering for a nonlinear Schrödinger equation with critical growth*, Selecta Math. (N.S.) **19** (2013), no. 2, 545–609, DOI 10.1007/s00029-012-0103-5. MR3090237

[4] Henri Berestycki and Thierry Cazenave, *Instabilité des états stationnaires dans les équations de Schrödinger et de Klein-Gordon non linéaires* (French, with English summary), C. R. Acad. Sci. Paris Sér. I Math. **293** (1981), no. 9, 489–492. MR646873

[5] H. Berestycki and P.-L. Lions, *Nonlinear scalar field equations. I. Existence of a ground state*, Arch. Rational Mech. Anal. **82** (1983), no. 4, 313–345, DOI 10.1007/BF00250555. MR695535

[6] Haïm Brezis and Elliott H. Lieb, *Minimum action solutions of some vector field equations*, Comm. Math. Phys. **96** (1984), no. 1, 97–113. MR765961

[7] Haïm Brézis and Louis Nirenberg, *Positive solutions of nonlinear elliptic equations involving critical Sobolev exponents*, Comm. Pure Appl. Math. **36** (1983), no. 4, 437–477, DOI 10.1002/cpa.3160360405. MR709644

[8] Luis A. Caffarelli, Basilis Gidas, and Joel Spruck, *Asymptotic symmetry and local behavior of semilinear elliptic equations with critical Sobolev growth*, Comm. Pure Appl. Math. **42** (1989), no. 3, 271–297, DOI 10.1002/cpa.3160420304. MR982351

[9] Shu-Ming Chang, Stephen Gustafson, Kenji Nakanishi, and Tai-Peng Tsai, *Spectra of linearized operators for NLS solitary waves*, SIAM J. Math. Anal. **39** (2007/08), no. 4, 1070–1111, DOI 10.1137/050648389. MR2368894

[10] Xing Cheng, Changxing Miao, and Lifeng Zhao, *Global well-posedness and scattering for nonlinear Schrödinger equations with combined nonlinearities in the radial case*, J. Differential Equations **261** (2016), no. 6, 2881–2934, DOI 10.1016/j.jde.2016.04.031. MR3527618

[11] Thomas Duyckaerts, Justin Holmer, and Svetlana Roudenko, *Scattering for the non-radial 3D cubic nonlinear Schrödinger equation*, Math. Res. Lett. **15** (2008), no. 6, 1233–1250, DOI 10.4310/MRL.2008.v15.n6.a13. MR2470397

[12] B. Gidas, Wei Ming Ni, and L. Nirenberg, *Symmetry of positive solutions of nonlinear elliptic equations in* $\mathbf{R}^n$, Mathematical analysis and applications, Part A, Adv. in Math. Suppl. Stud., vol. 7, Academic Press, New York-London, 1981, pp. 369–402. MR634248

[13] Manoussos Grillakis, *Linearized instability for nonlinear Schrödinger and Klein-Gordon equations*, Comm. Pure Appl. Math. **41** (1988), no. 6, 747–774, DOI 10.1002/cpa.3160410602. MR948770

[14] Justin Holmer and Svetlana Roudenko, *A sharp condition for scattering of the radial 3D cubic nonlinear Schrödinger equation*, Comm. Math. Phys. **282** (2008), no. 2, 435–467, DOI 10.1007/s00220-008-0529-y. MR2421484

[15] Slim Ibrahim, Nader Masmoudi, and Kenji Nakanishi, *Scattering threshold for the focusing nonlinear Klein-Gordon equation*, Anal. PDE **4** (2011), no. 3, 405–460, DOI 10.2140/apde.2011.4.405. MR2872122

[16] Reika Fukuizumi, *Remarks on the stable standing waves for nonlinear Schrödinger equations with double power nonlinearity*, Adv. Math. Sci. Appl. **13** (2003), no. 2, 549–564. MR2029931

[17] Yoshitsugu Kabeya and Kazunaga Tanaka, *Uniqueness of positive radial solutions of semilinear elliptic equations in* $\mathbf{R}^N$ *and Séré's non-degeneracy condition*, Comm. Partial Differential Equations **24** (1999), no. 3-4, 563–598, DOI 10.1080/03605309908821434. MR1683050

[18] Carlos E. Kenig and Frank Merle, *Global well-posedness, scattering and blow-up for the energy-critical, focusing, non-linear Schrödinger equation in the radial case*, Invent. Math. **166** (2006), no. 3, 645–675, DOI 10.1007/s00222-006-0011-4. MR2257393

[19] Sahbi Keraani, *On the defect of compactness for the Strichartz estimates of the Schrödinger equations*, J. Differential Equations **175** (2001), no. 2, 353–392, DOI 10.1006/jdeq.2000.3951. MR1855973

[20] Rowan Killip, Tadahiro Oh, Oana Pocovnicu, and Monica Vişan, *Solitons and scattering for the cubic-quintic nonlinear Schrödinger equation on* $\mathbb{R}^3$, Arch. Ration. Mech. Anal. **225** (2017), no. 1, 469–548, DOI 10.1007/s00205-017-1109-0. MR3634031

[21] Rowan Killip and Monica Visan, *The focusing energy-critical nonlinear Schrödinger equation in dimensions five and higher*, Amer. J. Math. **132** (2010), no. 2, 361–424, DOI 10.1353/ajm.0.0107. MR2654778

[22] Rowan Killip and Monica Vişan, *Nonlinear Schrödinger equations at critical regularity*, Evolution equations, Clay Math. Proc., vol. 17, Amer. Math. Soc., Providence, RI, 2013, pp. 325–437. MR3098643

[23] Elliott H. Lieb and Michael Loss, *Analysis*, 2nd ed., Graduate Studies in Mathematics, vol. 14, American Mathematical Society, Providence, RI, 2001, DOI 10.1090/gsm/014. MR1817225

[24] K. Nakanishi and W. Schlag, *Global dynamics above the ground state energy for the focusing nonlinear Klein-Gordon equation*, J. Differential Equations **250** (2011), no. 5, 2299–2333, DOI 10.1016/j.jde.2010.10.027. MR2756065

[25] K. Nakanishi and W. Schlag, *Global dynamics above the ground state energy for the cubic NLS equation in 3D*, Calc. Var. Partial Differential Equations **44** (2012), no. 1-2, 1–45, DOI 10.1007/s00526-011-0424-9. MR2898769

[26] Wei-Ming Ni and Izumi Takagi, *Locating the peaks of least-energy solutions to a semilinear Neumann problem*, Duke Math. J. **70** (1993), no. 2, 247–281, DOI 10.1215/S0012-7094-93-07004-4. MR1219814

[27] Jalal Shatah and Walter Strauss, *Instability of nonlinear bound states*, Comm. Math. Phys. **100** (1985), no. 2, 173–190. MR804458

[28] Terence Tao and Monica Visan, *Stability of energy-critical nonlinear Schrödinger equations in high dimensions*, Electron. J. Differential Equations (2005), No. 118, 28. MR2174550

[29] Terence Tao, Monica Visan, and Xiaoyi Zhang, *The nonlinear Schrödinger equation with combined power-type nonlinearities*, Comm. Partial Differential Equations **32** (2007), no. 7-9, 1281–1343, DOI 10.1080/03605300701588805. MR2354495

[30] Michael I. Weinstein, *Nonlinear Schrödinger equations and sharp interpolation estimates*, Comm. Math. Phys. **87** (1982/83), no. 4, 567–576. MR691044

[31] Michael I. Weinstein, *Modulational stability of ground states of nonlinear Schrödinger equations*, SIAM J. Math. Anal. **16** (1985), no. 3, 472–491, DOI 10.1137/0516034. MR783974

[32] Jian Zhang and WenMing Zou, *The critical case for a Berestycki-Lions theorem*, Sci. China Math. **57** (2014), no. 3, 541–554, DOI 10.1007/s11425-013-4687-9. MR3166237

Editorial Information

To be published in the *Memoirs*, a paper must be correct, new, nontrivial, and significant. Further, it must be well written and of interest to a substantial number of mathematicians. Piecemeal results, such as an inconclusive step toward an unproved major theorem or a minor variation on a known result, are in general not acceptable for publication.

Papers appearing in *Memoirs* are generally at least 80 and not more than 200 published pages in length. Papers less than 80 or more than 200 published pages require the approval of the Managing Editor of the Transactions/Memoirs Editorial Board. Published pages are the same size as those generated in the style files provided for $\mathcal{A}\mathcal{M}\mathcal{S}$-LaTeX or $\mathcal{A}\mathcal{M}\mathcal{S}$-TeX.

Information on the backlog for this journal can be found on the AMS website starting from `http://www.ams.org/memo`.

A Consent to Publish is required before we can begin processing your paper. After a paper is accepted for publication, the Providence office will send a Consent to Publish and Copyright Agreement to all authors of the paper. By submitting a paper to the *Memoirs*, authors certify that the results have not been submitted to nor are they under consideration for publication by another journal, conference proceedings, or similar publication.

Information for Authors

Memoirs is an author-prepared publication. Once formatted for print and on-line publication, articles will be published as is with the addition of AMS-prepared frontmatter and backmatter. Articles are not copyedited; however, confirmation copy will be sent to the authors.

Initial submission. The AMS uses Centralized Manuscript Processing for initial submissions. Authors should submit a PDF file using the Initial Manuscript Submission form found at `www.ams.org/submission/memo`, or send one copy of the manuscript to the following address: Centralized Manuscript Processing, MEMOIRS OF THE AMS, 201 Charles Street, Providence, RI 02904-2294 USA. If a paper copy is being forwarded to the AMS, indicate that it is for *Memoirs* and include the name of the corresponding author, contact information such as email address or mailing address, and the name of an appropriate Editor to review the paper (see the list of Editors below).

The paper must contain a *descriptive title* and an *abstract* that summarizes the article in language suitable for workers in the general field (algebra, analysis, etc.). The *descriptive title* should be short, but informative; useless or vague phrases such as "some remarks about" or "concerning" should be avoided. The *abstract* should be at least one complete sentence, and at most 300 words. Included with the footnotes to the paper should be the 2010 *Mathematics Subject Classification* representing the primary and secondary subjects of the article. The classifications are accessible from `www.ams.org/msc/`. The Mathematics Subject Classification footnote may be followed by a list of *key words and phrases* describing the subject matter of the article and taken from it. Journal abbreviations used in bibliographies are listed in the latest *Mathematical Reviews* annual index. The series abbreviations are also accessible from `www.ams.org/msnhtml/serials.pdf`. To help in preparing and verifying references, the AMS offers MR Lookup, a Reference Tool for Linking, at `www.ams.org/mrlookup/`.

Electronically prepared manuscripts. The AMS encourages electronically prepared manuscripts, with a strong preference for $\mathcal{A}\mathcal{M}\mathcal{S}$-LaTeX. To this end, the Society has prepared $\mathcal{A}\mathcal{M}\mathcal{S}$-LaTeX author packages for each AMS publication. Author packages include instructions for preparing electronic manuscripts, samples, and a style file that generates the particular design specifications of that publication series. Though $\mathcal{A}\mathcal{M}\mathcal{S}$-LaTeX is the highly preferred format of TeX, author packages are also available in $\mathcal{A}\mathcal{M}\mathcal{S}$-TeX.

Authors may retrieve an author package for *Memoirs of the AMS* from `www.ams.org/journals/memo/memoauthorpac.html`. The *AMS Author Handbook* is available in PDF format from the author package link. The author package can also be obtained free of charge by sending email to tech-support@ams.org or from the Publication Division, American Mathematical Society, 201 Charles St., Providence, RI 02904-2294, USA. When requesting an author package, please specify $\mathcal{A}_{\mathcal{M}}\mathcal{S}$-LaTeX or $\mathcal{A}_{\mathcal{M}}\mathcal{S}$-TeX and the publication in which your paper will appear. Please be sure to include your complete mailing address.

After acceptance. The source files for the final version of the electronic manuscript should be sent to the Providence office immediately after the paper has been accepted for publication. The author should also submit a PDF of the final version of the paper to the editor, who will forward a copy to the Providence office.

Accepted electronically prepared files can be submitted via the web at www.ams.org/submit-book-journal/, sent via FTP, or sent on CD to the Electronic Prepress Department, American Mathematical Society, 201 Charles Street, Providence, RI 02904-2294 USA. TeX source files and graphic files can be transferred over the Internet by FTP to the Internet node ftp.ams.org (130.44.1.100). When sending a manuscript electronically via CD, please be sure to include a message indicating that the paper is for the *Memoirs*.

Electronic graphics. Comprehensive instructions on preparing graphics are available at www.ams.org/authors/journals.html. A few of the major requirements are given here.

Submit files for graphics as EPS (Encapsulated PostScript) files. This includes graphics originated via a graphics application as well as scanned photographs or other computer-generated images. If this is not possible, TIFF files are acceptable as long as they can be opened in Adobe Photoshop or Illustrator.

Authors using graphics packages for the creation of electronic art should also avoid the use of any lines thinner than 0.5 points in width. Many graphics packages allow the user to specify a "hairline" for a very thin line. Hairlines often look acceptable when proofed on a typical laser printer. However, when produced on a high-resolution laser imagesetter, hairlines become nearly invisible and will be lost entirely in the final printing process.

Screens should be set to values between 15% and 85%. Screens which fall outside of this range are too light or too dark to print correctly. Variations of screens within a graphic should be no less than 10%.

Any graphics created in color will be rendered in grayscale for the printed version unless color printing is authorized by the Managing Editor and the Publisher. In general, color graphics will appear in color in the online version.

Inquiries. Any inquiries concerning a paper that has been accepted for publication should be sent to memo-query@ams.org or directly to the Electronic Prepress Department, American Mathematical Society, 201 Charles St., Providence, RI 02904-2294 USA.

Editors

This journal is designed particularly for long research papers, normally at least 80 pages in length, and groups of cognate papers in pure and applied mathematics. Papers intended for publication in the *Memoirs* should be addressed to one of the following editors. The AMS uses Centralized Manuscript Processing for initial submissions to AMS journals. Authors should follow instructions listed on the Initial Submission page found at www.ams.org/memo/memosubmit.html.

Selected Published Titles in This Series

1323 **Shu Kawaguchi and Kazuhiko Yamaki,** Effective Faithful Tropicalizations Associated to Linear Systems on Curves, 2021

1322 **D. Bulacu and B. Torrecillas,** Galois and Cleft Monoidal Cowreaths. Applications, 2021

1321 **Christian Haase, Andreas Paffenholz, Lindsey C. Piechnik, and Francisco Santos,** Existence of Unimodular Triangulations–Positive Results, 2021

1320 **Th. Heidersdorf and R. Weissauer,** Cohomological Tensor Functors on Representations of the General Linear Supergroup, 2021

1319 **Abed Bounemoura and Jacques Féjoz,** Hamiltonian Perturbation Theory for Ultra-Differentiable Functions, 2021

1318 **Chao Wang, Zhifei Zhang, Weiren Zhao, and Yunrui Zheng,** Local Well-Posedness and Break-Down Criterion of the Incompressible Euler Equations with Free Boundary, 2021

1317 **Eric M. Rains and S. Ole Warnaar,** Bounded Littlewood Identities, 2021

1316 **Ulrich Bunke and David Gepner,** Differential Function Spectra, the Differential Becker-Gottlieb Transfer, and Applications to Differential Algebraic K-Theory, 2021

1315 **S. Grivaux, É. Matheron, and Q. Menet,** Linear Dynamical Systems on Hilbert Spaces: Typical Properties and Explicit Examples, 2021

1314 **Pierre Albin, Frédéric Rochon, and David Sher,** Resolvent, Heat Kernel, and Torsion under Degeneration to Fibered Cusps, 2021

1313 **Paul Godin,** The 2D Compressible Euler Equations in Bounded Impermeable Domains with Corners, 2021

1312 **Patrick Delorme, Pascale Harinck, and Yiannis Sakellaridis,** Paley-Wiener Theorems for a p-Adic Spherical Variety, 2021

1311 **Lyudmila Korobenko, Cristian Rios, Eric Sawyer, and Ruipeng Shen,** Local Boundedness, Maximum Principles, and Continuity of Solutions to Infinitely Degenerate Elliptic Equations with Rough Coefficients, 2021

1310 **Hiroshi Iritani, Todor Milanov, Yongbin Ruan, and Yefeng Shen,** Gromov-Witten Theory of Quotients of Fermat Calabi-Yau Varieties, 2021

1309 **Jérémie Chalopin, Victor Chepoi, Hiroshi Hirai, and Damian Osajda,** Weakly Modular Graphs and Nonpositive Curvature, 2020

1308 **Christopher L. Douglas, Christopher Schommer-Pries, and Noah Snyder,** Dualizable Tensor Categories, 2020

1307 **Adam R. Thomas,** The Irreducible Subgroups of Exceptional Algebraic Groups, 2020

1306 **Kazuyuki Hatada,** Hecke Operators and Systems of Eigenvalues on Siegel Cusp Forms, 2020

1305 **Bogdan Ion and Siddhartha Sahi,** Double Affine Hecke Algebras and Congruence Groups, 2020

1304 **Matthias Fischmann, Andreas Juhl, and Petr Somberg,** Conformal Symmetry Breaking Differential Operators on Differential Forms, 2020

1303 **Zhi Qi,** Theory of Fundamental Bessel Functions of High Rank, 2020

1302 **Paul M. N. Feehan and Manousos Maridakis,** Łojasiewicz-Simon Gradient Inequalities for Coupled Yang-Mills Energy Functionals, 2020

1301 **Joachim Krieger,** On Stability of Type II Blow Up for the Critical Nonlinear Wave Equation in $\mathbb{R}^{3+1}$, 2020

1300 **Camille Male,** Traffic Distributions and Independence: Permutation Invariant Random Matrices and the Three Notions of Independence, 2020